Springer Tracts in Advanced Robotics

Volume 1

Springer

Berlin
Heidelberg
New York
Hong Kong
London
Milan
Paris
Tokyo

Fabrizio Caccavale · Luigi Villani (Eds.)

Fault Diagnosis and Fault Tolerance for Mechatronic Systems: Recent Advances

With 71 Figures

Springer

Professor Bruno Siciliano, Dipartimento di Informatica e Sistemistica, Università degli Studi di Napoli Federico II, Via Claudio 21, 80125 Napoli, Italy, email: siciliano@unina.it

Professor Oussama Khatib, Robotics Laboratory, Department of Computer Science, Stanford University, Stanford, CA 94305-9010, USA, email: khatib@cs.stanford.edu

Professor Frans Groen, Department of Computer Science, Universiteit van Amsterdam, Kruislaan 403, 1098 SJ Amsterdam, The Netherlands, email: groen@science.uva.nl

STAR (Springer Tracts in Advanced Robotics) has been promoted under the auspices of EURON (European Robotics Research Network)

Editors

Professor Fabrizio Caccavale
Dipartimento di Ingegneria e Fisica dell'Ambiente
Università degli Studi della Basilicata
Contrada Macchia Romana
85100 Potenza, Italy
e-mail: caccavale@unibas.it

Professor Luigi Villani
Dipartimento di Informatica e Sistemistica
Università degli Studi di Napoli Federico II
Via Claudio 21
80125 Napoli, Italy
e-mail: lvillani@unina.it

Cataloging-in-Publication Data applied for
Die Deutsche Bibliothek – CIP-Einheitsaufnahme
Fault diagnosis and fault tolerance for mechatronic systems: Recent advances / Fabrizio Caccavale ; Luigi Villani (ed.). - Berlin ; Heidelberg ; New York ; Hong Kong ; London ; Mailand ; Paris ; Tokio : Springer, 2003

(Springer tracts in advanced robotics ; Vol. 1)
(Engineering online library)

ISBN 3-540-44159-X Springer-Verlag Berlin Heidelberg New York

Springer-Verlag Berlin Heidelberg New York
a member of BertelsmannSpringer Science + Business Media GmbH

http://www.springer.de

Typesetting: Digital data supplied by author. Data-conversion by PTP-Berlin, Stefan Sossna e.K.
Cover-Design: design & production GmbH, Heidelberg
Printed on acid-free paper SPIN 10891275 62/3020Rw - 5 4 3 2 1 0

Editorial Advisory Board

EUROPE

Herman Bruyninckx, KU Leuven, Belgium
Raja Chatila, LAAS, France
Henrik Christensen, KTH, Sweden
Paolo Dario, Scuola S. Anna Pisa, Italy
Rüdiger Dillmann, Universität Karlsruhe, Germany

AMERICA

Ken Goldberg, UC Berkeley, USA
John Hollerbach, University of Utah, USA
Lydia Kavraki, Rice University, USA
Tim Salcudean, University of British Columbia, Canada
Sebastian Thrun, Carnegie Mellon University, USA

ASIA/OCEANIA

Peter Corke, CSIRO, Australia
Makoto Kaneko, Hiroshima University, Japan
Sukhan Lee, Samsung Advanced Institute of Technology, Korea
Yangsheng Xu, Chinese University of Hong Kong, PRC
Shin'ichi Yuta, Tsukuba University, Japan

Foreword

The field of robotics continues to flourish and develop. In common with general scientific investigation, new ideas and implementations emerge quite spontaneously and these are discussed, used, discarded or subsumed at conferences, in the reference journals, as well as through the Internet. After a little more maturity has been acquired by the new concepts, then archival publication as a scientific or engineering monograph may occur.

The goal of the *Springer Tracts in Advanced Robotics* is to publish new developments and advances in the fields of robotics research – rapidly and informally but with a high quality. It is hoped that prospective authors will welcome the opportunity to publish a structured presentation of some of the emerging robotics methodologies and technologies.

The edited volume by Fabrizio Caccavale and Luigi Villani concerns with a wide subject of modern engineering, namely mechatronics. This new discipline has lately been receiving an increasing deal of attention for its unique connotation to represent a blend of mechanical, electronic and information technologies. In such a scenario, the importance of providing the supervisory control system with fault detection and fault identification capabilities becomes crucial to the effective development of mechatronic systems.

Remarkably, besides the theoretical advancement in the field, the contributions are not restricted to robotic systems but they cover combustion, thermal, hydraulic systems and even outreach to underwater vehicles and aircrafts. A fine start for the series!

Napoli, July 2002 *Bruno Siciliano*

Foreword

[illegible]

Preface

This book is focused on the state-of-the-art of fault diagnosis for mechatronic systems. Beyond its survey aspects, the book aims at presenting challenging applications to various mechatronic systems, as well as relevant theoretical findings.

The contents reflects the lectures given by distinguished scholars at the workshop *Fault Diagnosis and Fault Tolerance for Dynamic Systems*, held in Vancouver, Canada, on October, 2002, in conjunction with the *IEEE International Symposium on Intelligent Control*. We asked each lecturer to extend her/his contribution to the workshop, thus giving birth to a chapter of the book.

The term "mechatronics" is used to designate the integration of mechanical, electronic and information technologies. Examples of mechatronic systems range from robotics and automotive systems to marine and aerospace vehicles. Often, mechatronic systems operate in remote or hazardous environments, where a high degree of autonomy and safety is required. On the other hand, in a wide class of applications, a close interaction with humans is of concern; in this case the main goal is to achieve a safe and reliable man-machine interaction. Therefore, designing of systems with self-diagnosis capabilities is becoming a crucial challenge in mechatronics.

A reading track along the six chapters of the book is briefly outlined in the following.

In mechatronic systems fast and accurate detection of anomalous situations (fault detection) as well as identification of their causes (fault identification) is of the utmost importance. This problem can be tackled by resorting to model-based fault diagnosis approaches, e.g., to observer-based techniques. The first chapter, co-authored by *M. Saif* and *Y. Xiong*, is focused on observer-based fault diagnosis. Namely, two different sliding mode observer design strategies are reviewed. Moreover, a new sliding mode observer for a class of nonlinear uncertain systems is proposed, and a number of mechatronic application examples is provided.

Whenever a deterministic model of the plant is not available, e.g., the system is subject to parameter randomness, fault diagnosis can be achieved by considering stochastic models. In the second chapter *H. Wang* addresses this problem with reference to non-Gaussian stochastic systems. Also, the problem of fault tolerant control is formulated and solved into an adaptive framework.

Embedding fault diagnosis in mechatronic systems is becoming critical to ensure higher levels of safety and reliability in automated factory plants and autonomous systems. A crucial component of automated factory plants is the industrial robot. The adoption of fault diagnosis techniques for industrial robots plays a central role in order to achieve a safe man-robot interaction as well as a quick and appropriate reaction of the robotic system to the occurrence of failures. In the third chapter, authored by *F. Caccavale* and *L. Villani*, different observer-based approaches to fault diagnosis for mechanical manipulators are presented and critically compared. All the considered schemes are experimentally tested on a six-degree-of-freedom industrial robot. An application to autonomous systems operating in remote environments is presented in the fourth chapter by *G. Antonelli*. Namely, several fault detection/tolerance strategies for autonomous underwater vehicles (AUVs) and remotely operated vehicles (ROVs) are surveyed.

An important application of mechatronics in aerospace systems is represented by flight controllers. In the fifth chapter, authored by *J.D. Bošković* and *K. Mehra*, issues in fault detection, identification and reconfiguration in flight control are addressed. Numerical simulations developed for a combat aircraft are carried out to illustrate the theoretical findings.

The last chapter, authored by *M.L. Leuschen*, *I.D. Walker* and *J.R. Cavallaro* is focused on the application of the analytical redundancy method to fault detection for electro-hydraulic systems. In order to cope with the typical nonlinear behavior of hydraulic systems, the analytical redundancy approach is generalized to nonlinear dynamical systems.

The book is intended for graduate students, researchers, scientists and scholars who wish to update their knowledge in fault diagnosis and fault tolerant control of mechatronic systems. We hope they will find the book useful.

Napoli,
June 2002

Fabrizio Caccavale
Luigi Villani

Contents

Sliding Mode Observers and Their Application in Fault Diagnosis

Mehrdad Saif and Yi Xiong

School of Engineering Science
Simon Fraser University
8888 University Drive
Vancouver, British Columbia V5A 1S6 CANADA

Abstract. Two commonly used approaches to sliding mode observer (SMO) design, namely the equivalent control approach of Utkin and the Walcott and Zak's observer design strategies are reviewed. Ceratin limitations of each design strategy are discussed and two alternative design approach for the Walcott and Zak observer based on the representation of a linear system in special coordinate basis (SCB) from are given. Additionally, a comparative discussion between the SMO and the unknown input observer (UIO) is provided, along with a discussion on similarities and differences between the two observer design strategies. Next a new sliding mode observer for linear uncertain systems is discussed. The advantage of this observer is that it can be built under much less conservative conditions than the one discussed. In addition, we address the issue of estimating a function of the state as well as unknown inputs or structural uncertainties. Furthermore, basic SMO design idea is extended to certain class of nonlinear uncertain systems. Next, we discuss how these SMOs can be used for fault detection and isolation (FDI) purposes. Finally, a number of examples illustrating the application of the SMO in mechatronic applications such as fault diagnosis of an internal combustion engine, robots, and electric motors are presented.

1 Introduction

System monitoring and timely fault detection capabilities are critical requirements of many modern mechatronic systems. Traditionally these features have been of utmost importance in safety critical systems such as civil and military aviation, or nuclear power plants, etc. However, in recent years, other factors have been playing a major role in recognizing the need for these capabilities in other technical systems. Broadly speaking, by the term *fault* we mean failures, errors, malfunctions or disturbances in the functional units that can lead to undesirable or intolerable behavior of the system. Some of the contributing factors that have made the automatic fault detection, isolation and accommodation (FDIA) problem to become an active area for research in a wide variety of industries and systems are:

i. The increased level of sophistication of many industrial and consumer goods due to the advances in electronics and computer technology, and

F. Caccavale, L. Villani (Eds.): Fault Diagnosis and Fault Tolerance for Mechatronic Systems, STAR 1, pp. 1-57, 2003.

at the same time decrease in processor's costs. Today's automobiles are a good example of this mechatronic trend. The auto manufacturers have introduced a tremendous amount of electronics in recent models. Many functions such as powertrain control, anti-lock brake, chassis control, climate control, traction control, etc. are now performed electrically and are available on many vehicles.

ii. Many manufacturing and process industries are highly interested in FDIA capabilities due to the fact that timely detection of early faults can result in unexpected and total failure that can lead to plant shutdown and loss of revenues. Therefore, economics is now an important factor in incorporation of FDIA techniques in many industries.

iii. The environmental concern is now a new driving force for FDI requirement. An example of this is the recent California Air Resource Board (CARB), and Environmental Protection Agency (EPA) legislations which required that as of 1998, On Board Diagnostics II (OBD-II) to be rolled into all light duty vehicles sold in North American fleet. Essentially, OBD-II requires fault detection capability for all vehicle components whose failure can result in emission levels beyond a certain level. It would not be surprising if similar tight restriction were to be placed on control and fault diagnosis of other internal combustion engines such as those of boats or lawn mowers.

In any of the systems that were discussed above, in order to have the efficient operation of the process and to increase the reliability and safety, prompt detection of anomalous situations (fault detection) and the fast identification (isolation) of the most probable causes (faults) need to be addressed. FDI can be carried out using analytical or functional information about the system being monitored, i.e., based on a mathematical model of the system. This approach is known as analytical redundancy, which is also known as model–based or quantitative FDI. Model–based FDI is currently the subject of extensive research and is being used in highly reliable control systems due to the fact that analytical redundancy based techniques are more economical and at times more powerful. These methods have the potential of detecting soft incipient faults even during the system's transient operation. One of the most popular of the model based approaches to FDI is that of the observer based techniques. The basic idea of observer-based methods consists of reconstruction of the outputs of the system of interest with the aid of observers or Kalman filters, and the use of the estimation error (or innovation, respectively) as the residual. The observer feedback gain enters the calculation of the residual generator and the gain design problem provides freedom for achieving the required performance (e.g. see [18],[45], [19]). However, special attention has to be paid when applying observer theory for fault detection and isolation. The basic configuration of observer-based fault diagnosis is shown in Figure 1.

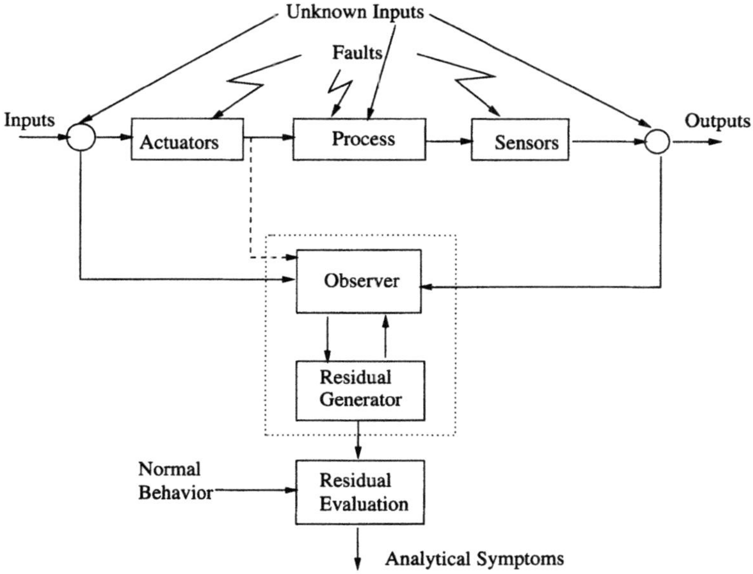

Fig. 1. Basic configuration of observer-based fault diagnosis

The early works were based on the assumption that the system under consideration is linear and that a sufficiently accurate mathematical model of the system is available. False alarm may frequently occur when these approaches are used to linear systems subject to unknown inputs, because the fault signal and unknown input are very likely to produce a similar residual signal. A straightforward method to create robustness with respect to unknown inputs is to generate unknown input decoupled residuals. If conditions for a perfect decoupling are not fulfilled, one can determine an optimal approximation in the sense of compromise between sensitivity with respect to faults, and robustness with respect to unknown inputs. The problem of perfect decoupling of faults from the unknown inputs has been attacked in both frequency and time domain. A frequency domain solution was given by Frank and Ding [17]. There are three kinds of time domain solutions. The first one is known as the *unknown input observer (UIO)* based approach (see Chang and Hsu [8], Chen *et al.* [9], Frank *et al.* [18], Saif and Guan [45]). The second is the eigenstructure assignment approach (Patton and Chen [38], Wang and Daley [57]). These two approaches do not make assumptions on the size and the time functions or on the frequency characteristics of the faults or of the unknown inputs. The third is the sliding mode FDI observer (Hermans and Zarrop [25]), where unknown inputs are assumed to be bounded. The practical importance of the decoupling approach lies in the fact that it allows small faults to be detected, even if there are large modeling errors. Of course, certain conditions have to be satisfied in order for decoupling between faults and unknown inputs to be made possible.

Recently, some methods to improve the robustness of Beard and Jones Detection Filter were proposed by [13], and [10]. The GLRT was generalized by [35] to make it insensitive to a large class of noise and plant model uncertainties. These works were based on the newly developed H_∞ and differential game theory.

The investigations mentioned above are based on linear multivariable uncertain systems with a nominally linear time-invariant part and an unknown input part. Many industrial processes are of nonlinear nature, and consequently, have a nonlinear mathematical model. Linear robust approaches are limited if the system to be monitored is strongly nonlinear. Linear uncertain system models can cover a small class of nonlinear systems by representing nonlinear parts as unknown inputs. However, they will introduce too many unknown inputs which will make perfect or approximate decoupling difficult. Therefore, the study of nonlinear observer-based FDI has received considerable attention in the past few years. Apart from the problem of nonlinear observer based FDI, the subject of nonlinear observer design in itself is very much an active area of current research. One nonlinear observer design approach deals with nonlinear systems for which observers with linearizable error dynamics can be designed (see e.g. [2], [30], [31], [41], [59]). In several works (e.g. [4], [6],[21], [39]) systems which are composed of a linear unforced part and a nonlinear state dependent controlled part are considered. In [22], an observer is given for a class of nonlinear systems which are not necessarily control affine. However, the gain of the proposed observer is not easily computable. Recently [7,27] proposed observers based on some ideas from the high gain approach whose gain could easily be designed.

With the application of nonlinear observer theory, some nonlinear system FDI approaches have been obtained, principally in the detection, and also with some restrictions, for isolation of faults. Based on a Thau-type nonlinear observer, the BJDF method is generalized in [20] for a class of Lipschitz nonlinear systems, with the linearity assumption being made on the output vector. A nonlinear observer, which was constructed by sliding mode design techniques, was used for fault diagnosis of control affine nonlinear systems in [34]. For state affine systems, Hammouri *et al.* used Kalman-like time-varying observers to build the residual for FDI [24]. The fault diagnosis based on UIO for linear systems (in different versions) is generalized to bilinear systems in [66], and [67,68]. Seliger and Frank [48], Yang and Saif [65] extended linear UIO to a more general class of nonlinear systems by applying a nonlinear state transformation, and applied their proposed nonlinear UIO to fault diagnosis for nonlinear uncertain systems.

Recently, a considerable number of papers have been devoted to the topic of observers design based on sliding mode concept [14,49,54–56,70,16,64]. Utkin presents a sliding mode observer strategy for linear systems whereby the error between the estimated and measured outputs is forced to exhibit a sliding mode and measurement noise effects are reduced [54]. Walcott and Zak use a Lyapunov-based approach to formulate an observer design which exhibits asymptotic state error decay in the presence of bounded nonlinearity/uncertainties that satisfy the so-called matching condition [55]. This observer is also called as the min-max observer due to its sliding mode properties [70]. Unlike several other techniques for observer design of nonlinear systems where the exact knowledge of the plant's nonlinearities must be known and either directly or indirectly incorporated into the dynamics of the observer, only the bounds of the nonlinearities of the plant are used in Walcott and Zak observer dynamics. Moreover, these dynamics may easily be implemented. In spite of these appealing properties, the design method of Walcott and Zak observer was impractical for high order systems until Edwards and Spurgeon proposed a numerically tractable algorithm [16]. Sreedhar and Fernandez explore the practical application of sliding mode observers for fault detection if full state measurements are available [52]. Hermans and Zarrop [25] investigate the potential and advantages of the sliding mode observers for robust fault detection based on the design procedure given in [16].

In [14] the approach of of designing an extended the sliding mode observer using equivalent control for linear systems [54] was extended to nonlinear systems of the form

$$\begin{aligned} \dot{x} &= f(x) + B(x)u \\ y &= h(x) \end{aligned} \tag{1}$$

This extension was also applied to nonlinear systems in triangular input form in [1], output and output derivative injection form in [5] and was applied to FDI problem in [12] . In papers [49] and [58], a framework similar to a Luenberger observer were used by appending a switching function with constant or time-varying gains.

In summary the existing SMO design methods can be classified in two categories: 1) The equivalent control based methods, and 2) sliding mode observer designs based on the method of Lyapunov. Since these techniques are important to the developments in this article, we shall briefly expand on these approaches in the following:

2 Sliding Mode Observers for Linear and Nonlinear Systems

In this section we shall review two popular SMO design strategies and propose new SMO designs based on these strategies.

2.1 SMO Design Based on Utkin's Method of Equivalent Control

Observers based on sliding mode approach first were developed for linear systems [54]. Consider a linear uncertain dynamic system described by

$$\begin{aligned} \dot{x} &= Ax + Bu + Gd(x,u,t) \\ y &= Cx \end{aligned} \tag{2}$$

where $x \in \mathcal{R}^n$ is the state, $u \in \mathcal{R}^m$ is the control input, $y \in \mathcal{R}^p$ is the output. The matrices A, B and C are of appropriate dimensions. It is assumed that $d(x,u,t)$ is unknown, but bounded, so that

$$\|d(x,u,t)\| \leq \rho \quad \forall x \in \mathcal{R}^n, u \in \mathcal{R}^m, t \geq 0$$

where $\|\cdot\|$ refers to the Euclidean norm. G is a full rank matrix in $\mathcal{R}^{n\times q}$. $Gd(x,u,t)$ represents the system uncertainties or nonlinearities, namely the unknown input. In addition, the matrices B and C are assumed to be of full rank. For the moment assume that $d = 0$ and the pair $\{A, C\}$ is observable. Furthermore, without the loss of generality, it can be assumed that the output distribution matrix can be written as

$$C = [C_1 \quad C_2]$$

where $C_1 \in \mathcal{R}^{p\times(n-p)}, C_2 \in \mathcal{R}^{p\times p}$ and $det(C_2) \neq 0$. Consequently, the transformation

$$T = \begin{bmatrix} I_{n-p} & 0 \\ C_1 & C_2 \end{bmatrix} \tag{3}$$

is non-singular and with respect to this new coordinate system it can be seen that the new output distribution matrix $CT^{-1} = [0 \quad I_p]$. If the other system matrices are written as

$$TAT^{-1} = \begin{bmatrix} A_{11} & A_{12} \\ A_{21} & A_{22} \end{bmatrix} \quad \text{and } TB = \begin{bmatrix} B_1 \\ B_2 \end{bmatrix}$$

then the nominal system can be written as

$$\begin{aligned} \dot{x}_1 &= A_{11}x_1 + A_{12}y + B_1u \\ \dot{y} &= A_{21}x_1 + A_{22}y + B_2u \end{aligned} \tag{4}$$

The corresponding sliding mode observer for the y subsystem is

$$\dot{\hat{y}} = A_{22}\hat{y} + A_{21}\hat{x}_1 + B_2u + L_1 sign(y - \hat{y}) \tag{5}$$

where $(\hat{x}_1, \hat{y})$ are the estimates of (x_1, y), L_1 is a constant nonsingular feedback gain matrix, and $sign(.)$ is the sign function. If the error between the estimates and the true states are written as $e_y = y - \hat{y}$ and $e_1 = x_1 - \hat{x}_1$, then the following error system is obtained:

$$\dot{e}_y = A_{22}e_y + A_{21}e_1 - L_1 sign(e_y). \tag{6}$$

It can be shown using singular perturbation theory that for a large enough L_1, a sliding mode motion can be induced on the outputs' error state in (6) [54]. It follows that, after some finite time t_s, for all subsequent time, $e_y = 0$ and $\dot{e}_y = 0$. For the second subsystem, the observer equation is

$$\dot{\hat{x}}_1 = A_{11}\hat{x}_1 + A_{12}y + B_1 u + L_2 L_1 sign(e_y) \tag{7}$$

which results in the following estimation error dynamics

$$\dot{e}_1 = A_{11}e_1 - L_2 L_1 sign(e_y). \tag{8}$$

According to the equivalent control method, the system in sliding mode behaves as if $L_1 sign(e_y)$ is replaced by its equivalent value $(L_1 sign(e_y))_{eq}$, which can be calculated from the subsystem (6) assuming $e_y = 0$ and $\dot{e}_y = 0$. Hence

$$(L_1 sign(e_y))_{eq} = A_{21}e_1. \tag{9}$$

Substituting (9) into (8) we obtain

$$\dot{e}_1 = (A_{11} - L_2 A_{21})e_1. \tag{10}$$

Note that the pair $\{A_{11}, A_{21}\}$ is observable if $\{A, C\}$ is observable, therefore through an appropriate choice of L_2 the estimation error e_1 can be made to approach zero as $t \to \infty$.

To see how the performance of the observer will be affected as a result of uncertainties in the system, assume now that the term Gd in (2) is no longer zero. Then using the transformation (3) for Utkin SMO, the linear uncertain system (2) can be transformed into following canonical form,

$$\begin{aligned} \dot{x}_1 &= A_{11}x_1 + A_{12}y + B_1 u + G_1 d(x,u,t) \\ \dot{y} &= A_{21}x_1 + A_{22}y + B_2 u + G_2 d(x,u,t) \end{aligned} \tag{11}$$

Under the sliding observer (5), the equivalent control signal will be

$$(L_1 sign(e_y))_{eq} = A_{21}e_1 + G_2 d(x,u,t). \tag{12}$$

The error dynamics of e_1 in this case will become,

$$\dot{e}_1 = (A_{11} - L_2 A_{21})e_1 + (G_1 + L_2 G_2)d(x,u,t). \tag{13}$$

Based on the above error dynamics, $e_1(t)$ will not in general approach zero if d is nonzero. Also, note that even if the first subsystem in (11) is unknown input free (i.e. $G_1 = 0$), the equivalent control signal will still introduce the uncertainties into its observer error dynamics. To have a convergent estimate one approach may be to select the gain L_2 such that $G_1 + L_2 G_2 = 0$. However, in general it may be very difficult to satisfy this added constraint on the observer gain L_2. A more reasonable approach to get around this difficulty is to force the estimation error to be below an acceptable threshold. This can be achieved through a suitable choice of the gain L_2 under the assumption that

$\|d(x,u,t)\|$ is small enough, as discussed in [49,58]. Obviously in such a case the observer's performance is not guaranteed if $\|d(x,u,t)\|$ is not small. The idea described above can be extended to nonlinear systems of the form (1). This extension is covered in [14], and was also applied to nonlinear systems in triangular input form in [1], output and output derivative injection form in [5], and general nonlinear observable systems with measurements that are linearly related to the state vector (i.e. $y = Cx$) in [49,58].

2.2 Walcott and Zak Sliding Mode Observer

The Walcott and Zak's sliding mode observer ([55,56]) which was further developed by Misawa [37] attempts to provide an exponentially convergent estimate of the state of the dynamical system described in (2) despite the presence of *matched* uncertainty. Note that the term $d(x,u,t)$ in (2) can represent not only system nonlinearities, but also time-varying terms or internal/external uncertainties/disturbances.

Recall that it was assumed that the pair $\{A, C\}$ is completely observable in which case a matrix $K \in \mathcal{R}^{n\times p}$ exists such that $A_0 = A - KC$ is Hurwitz. Therefore, for every real symmetrical positive definite (SPD) matrix $Q \in \mathcal{R}^{n\times n}$, there exists a real SPD matrix P as the unique solution to the following Lyapunov equation:

$$P(A-KC) + (A-KC)^T P = -Q \tag{14}$$

It is also assumed that a Lyapunov pair $\{P, Q\}$ for A_0 exists such that the structural constraint

$$WC = G^T P \tag{15}$$

is satisfied for some $W \in \mathcal{R}^{q\times p}$. The proposed observer in [55] has the form

$$\dot{\hat{x}} = A\hat{x} + Bu + K(y - x) + \nu \tag{16}$$

where

$$\nu = \begin{cases} \rho\frac{GWCe}{\|WCe\|} & \text{if } WCe \neq 0 \\ 0 & \text{otherwise} \end{cases} \tag{17}$$

or

$$\nu = \rho G sign(WCe) \tag{18}$$

where $e = x - \hat{x}$, and $sign(\cdot)$ is the sign function. Note that this is basically a conventional Luenberger observer with the additional nonlinear discontinuous "switching" term ν. By using $V(e) = e^T Pe$ as a Lyapunov function candidate it can be shown that $\dot{V}(e) < 0$ for $e \neq 0$ and thus $e \to 0$ exponentially, and the surface $S = \{e : WCe = 0\}$ is the sliding manifold. The main difficulty in designing the Walcott and Zak observer is that of computing the matrix pair $\{P, W\}$ such that (14) and (15) are satisfied. This is not an easy task because P is a Lyapunov matrix for A_0, where A_0 itself is dependent on the choice of K. Nevertheless, the Walcott and Zak [56] design algorithm can be summarized as follows:

i. Choose the spectrum of A_0, and compute K accordingly;
ii. Solve the structural constraint *symbolically* to obtain an expression for P in terms of entries of W, ensuring that P is symmetric;
iii. Compute Q symbolically in terms of entries of P using the expression $-(PA_0 + A_0^T P)$;
iv. Choose the elements of W and P to ensure Q is SPD.

The Walcott and Zak observer is very simple in its structure and the notion of total insensitivity of the observer to the matched uncertainties is appealing. However, from a computational point of view, this framework is impractical for high order systems because of the manipulation and solution of the associated constrained Lyapunov problem defined by equations (14) and (15). Furthermore, it is difficult to verify the existence of the Walcott-Zak observer for systems (A, G, C) based on the original system matrices. In [51] it is shown that a sufficient condition for(15) being valid is that the modified transfer function defined as $G_F(s) = WC(sI - A_0)^{-1}G$ be strictly positive real. However, [51] did not consider the problem of finding a suitable W. Edwards and Spurgeon [16] provided a numerically tractable algorithm by transforming the system (A, G, C) into a canonical form.

In the next subsection an alternative approach for designing the Walcott and Zak observer based on the representation of the system in special coordinate basis (SCB) form is discussed. Further the SCB form will lay the foundation for the remainder of the discussions in this chapter.

2.3 Special Coordinate Basis (SCB) Form

First consider some preliminary results from the SCB form of linear systems given below.

Theorem 1. *[42] For system (2), there exist non-singular transformations Γ_1, Γ_2 and Γ_3, and integer $m_d \leq q, q_i = 1, ..., m_d$, such that*

$$\overline{x} = \Gamma_1^{-1}x, \overline{y} = \Gamma_2^{-1}y, \overline{d} = \Gamma_3^{-1}d;$$

$$\overline{x} = [(x_a)^T, x_b^T, x_c^T, x_d^T]^T, x_a = [(x_a^-)^T, (x_a^+)^T]^T$$

$$\overline{y} = [y_d^T, y_b^T]^T, \overline{d} = [d_d^T, d_c^T]^T$$

$$y_d = [y_1, y_2, ..., y_{m_d}]^T, d_d = [d_1, d_2, ..., d_{m_d}]^T$$

and

$$\begin{aligned}
\dot{x}_a &= A_a x_a + L_{ad} y_d + L_{ab} y_b + B_a u \\
\dot{x}_b &= A_b x_b + L_{bd} y_d + B_b u, \qquad y_b = C_b x_b \\
\dot{x}_c &= A_c x_c + L_{cd} y_d + L_{cb} y_b + G_c E_{ca} x_a + G_c d_c + B_c u
\end{aligned} \tag{19}$$

and $x_d = \left[x_{1d}\ x_{2d}\ ...\ x_{m_d d}\right]^T$, *for each* $x_{id}, i = 1, ..., , m_d$,

$$\dot{x}_{id}=A_{q_i}x_{id}+L_{id}y_d+G_{q_i}[E_{ia}x_a+E_{ib}x_b+E_{ic}x_c+\sum_{j=1}^{m_d}E_{ij}x_{jd}+d_i]+B_{id}u \quad (20)$$

$$y_i = C_{q_i}x_{id}, \qquad y_d = C_d x_d \quad (21)$$

Here the states x_a, x_b, x_c, x_d *are of dimensions* n_a, n_b, n_c *and* $n_d = \sum_{i=1}^{m_d} q_i$ *respectively, while* x_{id} *is of dimension* q_i *for each* $i = 1, ..., m_d$. *The matrices* A_{q_i}, G_{q_i} *and* C_{q_i} *have following form:*

$$A_{q_i} = \begin{bmatrix} 0 & I_{q_i-1} \\ 0 & 0 \end{bmatrix}; G_{q_i} = \begin{bmatrix} 0_{(q_i-1)\times 1} \\ 1 \end{bmatrix}; C_{q_i} = \begin{bmatrix} 1 & 0_{1\times(q_i-1)} \end{bmatrix}; \quad (22)$$

and last row of L_{id} *is identically zero.*

To this end, x_d subsystem is further decomposed into m_d subsystems. Let

$$x_{id} = \left[x_1^i\ x_2^i\ ...\ x_{q_i}^i\right]^T;$$

The special form of A_{q_i}, G_{q_i} implies that the equations of subsystem x_{id} in (20) can be rewritten as

$$\begin{array}{ll} \dot{x}_1^i & = x_2^i + L_1^i y_d + B_1^i u \\ \dot{x}_2^i & = x_3^i + L_2^i y_d + B_2^i u \\ \cdots & \cdots \\ \dot{x}_{q_i-1}^i & = x_{q_i}^i + L_{q_i-1}^i y_d + B_{q_i-1}^i u \\ \dot{x}_{q_i}^i & = E_{ia}x_a + E_{ib}x_b + E_{ic}x_c + \sum_{j=1}^{m_d} E_{ij}x_{jd} + B_{q_i}^i u + d_i \\ y_i & = x_1^i \end{array} \quad (23)$$

In a more compact notation the above representation of the system in the SCB form can be written as:

$$\overline{\Sigma}: \begin{cases} \dot{\overline{x}} = \overline{A}\overline{x} + \overline{B}u + \overline{G}\overline{d} \\ \overline{y} = \overline{C}\overline{x} \end{cases}$$

where $(\overline{A}, \overline{G}, \overline{C})$ are given by

$$\overline{A} := \Gamma_1^{-1} A \Gamma_1 = \begin{bmatrix} A_{aa}^- & 0 & L_{ab}^- C_b & 0 & L_{ad}^- C_d \\ 0 & A_{aa}^+ & L_{ab}^+ C_b & 0 & L_{ad}^+ C_d \\ 0 & 0 & A_b & 0 & L_{bd} C_d \\ G_c E_{ca}^- & G_c E_{ca}^+ & L_{cb} C_b & A_c & L_{cd} C_d \\ G_d E_a^- & G_d E_a^+ & G_d E_b & G_d E_c & A_d \end{bmatrix}$$

$$\overline{G} := \Gamma_1^{-1} G \Gamma_3 = \begin{bmatrix} 0 & 0 \\ 0 & 0 \\ 0 & 0 \\ 0 & G_c \\ G_d & 0 \end{bmatrix}, \overline{C} := \Gamma_2^{-1} C \Gamma_1 = \begin{bmatrix} 0 & 0 & 0 & 0 & C_d \\ 0 & 0 & C_b & 0 & 0 \end{bmatrix} \quad (24)$$

where

$$A_{aa} = \begin{bmatrix} A_{aa}^- & 0 \\ 0 & A_{aa}^+ \end{bmatrix}$$

$$E_a = [E_a^-, E_a^+], E_{ca} = [E_{ca}^-, E_{ca}^+].$$

and G_d, C_d is in the form of

$$G_d = \begin{bmatrix} G_{q1} & 0 & \dots & 0 \\ 0 & G_{q2} & \dots & 0 \\ \dots & \dots & \dots & 0 \\ 0 & 0 & 0 & G_{qd} \end{bmatrix}; C_d = \begin{bmatrix} C_{q1} & 0 & \dots & 0 \\ 0 & C_{q2} & \dots & 0 \\ \dots & \dots & \dots & 0 \\ 0 & 0 & 0 & C_{qd} \end{bmatrix}; \tag{25}$$

The matrices G_{qi}, C_{qi} have the following form:

$$G_{qi} = \begin{bmatrix} 0 \\ 1 \end{bmatrix}; C_{qi} = \begin{bmatrix} 1 & 0 \end{bmatrix}; \tag{26}$$

Obviously for the case $q_i = 1$, we have $G_{qi} = 1, C_{qi} = 1$.

Some of the important properties of the SCB which are relevant to our discussion are outlined below [42]:

Property 1 The system Σ is right invertible if and only if x_b and hence y_b do not exist (i.e. $n_b = 0$), and left invertible if and only if x_c and hence u_c do not exist (i.e. $n_c = 0$). Finally, the system is right invertible if and only if it is both right and left invertible.

Property 2 The invariant zeros of the systems are the eigenvalues of A_a. If A_a is further decomposed, the number of stable and unstable zeros of the system are given by n_a^-, n_a^+ respectively . Eigenvalues of A_{aa}^- and A_{aa}^+ are stable and unstable zeros of the system respectively.

Property 3 n_d is the number of infinite zeros, and $n_d = \sum_{i=1}^{m_d} q_i$, where $q_i, i = 1, ..., m_d$ is the number of infinite zeros of order of i, m_d is the highest order of an infinite zero. Further, $m_d = rank(C_d) = rank(G_d)$.

Property 4 (A_b, C_b) and (A_d, C_d) form observable pairs. The subsystems involving x_a and x_c are unobservable. The pair $\{A, C\}$ is observable (detectable) if and only if $\{A_{ob}, C_{ob}\}$ is an observable (detectable) pair, where

$$A_{ob} = \begin{bmatrix} A_{aa} & 0 \\ G_c E_{ca} & A_{cc} \end{bmatrix}, C_{ob} = \begin{bmatrix} E_a & E_{ca} \end{bmatrix}$$

As will be seen in this section, the SCB form of the system plays a dominant role also in our Walcott and Zak observer analysis and synthesis.

2.4 Lyapunov Equation Based Walcott and Zak's SMO Design

In this section we shall introduce a systematic approach for Walcott and Zak's SMO design based on the constrained Lyapunov equation discussed before.

Lemma 1. *Let $(\overline{A}, \overline{G}, \overline{C})$ be SCB form of the original system (A, G, C). $(\overline{A}, \overline{C})$ is observable if and only if (A, C) is observable. Furthermore, $P = \Gamma_1^{-T}\overline{P}\Gamma_1^{-1}, Q = \Gamma_1^{-T}\overline{Q}\Gamma_1^{-1}$ is a Lyapunov matrix pair for $A_0 = A - KC$ which satisfies the structural constraint $WC = G^T P$, where $K = \Gamma_1 \overline{K}\Gamma_2^{-1}$, $W = \Gamma_3^{-T}\overline{W}\Gamma_2^{-1}$ if and only if there exists $\overline{K}$ such that a Lyapunov matrix pair $\{\overline{P}, \overline{Q}\}$ for $\overline{A}_0 = \overline{A} - \overline{KC}$ satisfies the structural constraint $\overline{WC} = \overline{G}^T\overline{P}$.*

Proof: By direct validation. ∎

Lemma 2. *Given matrices P, C, and G, there exists W satisfying $WC = G^T P$ if and only if*

$$G^T P(I - C^\dagger C) = 0$$

in which case a general solution is given by

$$W = G^T PC^\dagger + Z - ZCC^\dagger \tag{27}$$

where Z is arbitrary matrix of the same size as W, and $C^\dagger$ is generalized inverse of C.

Proof: It is a straightforward application of the fundamental result of linear matrix equation $AXB = Y$, where A, B, and Y are given with adequate dimensions, and X is the unknown. ∎

Lemma 3. *[26] Suppose a symmetric matrix P is partitioned as*

$$\begin{bmatrix} P_{11} & P_{12} \\ P_{12}^T & P_{22} \end{bmatrix}$$

where P_{11}, and P_{22} are square matrices. The matrix P is positive definite if and only if P_{11} is positive definite and $P_{22} > P_{12}^T P_{11}^{-1} P_{12}$. Furthermore, this condition is equivalent to having $\rho(P_{12}^T P_{11}^{-1} P_{12} P_{22}^{-1}) < 1$, where $\rho(.)$ represents spectral radius.

Theorem 2. *If the pair $\{A, C\}$ is completely observable, a Walcott and Zak observer (16) for system (2) exists if and only if the system (A, G, C) is minimum phase and $rank(CG) = rank(G)$.*

Proof: Based on Lemma 2, it is assume that the (A, G, C) is already in SCB form.

Necessity. Let K be any appropriate gain matrix so that $A_0 = A - KC$ is stable with an associated Lyapunov matrix P such that there exists a W satisfying

$$WC = G^T P$$

Let $P = M^T M$, where M is nonsingular. Following Lemma 3, then

$$G^T P(I - C^\dagger C) = G^T M^T M(I - C^\dagger C) = 0 \tag{28}$$

From (24),

$$I - C^\dagger C = \begin{bmatrix} I_{n_a^-} & 0 & 0 & 0 & 0 \\ 0 & I_{n_a^+} & 0 & 0 & 0 \\ 0 & 0 & I_{n_b} - C_b^\dagger C_b & 0 & 0 \\ 0 & 0 & 0 & I_{n_c} & 0 \\ 0 & 0 & 0 & 0 & I_{n_d} - C_d^\dagger C_d \end{bmatrix} \tag{29}$$

Let $\hat{I}_b = I_{n_b} - C_b^\dagger C_b, \hat{I}_d = I_{n_d} - C_d^\dagger C_d$. By partitioning M as

$$M = \begin{bmatrix} M_a^- & M_a^+ & M_b & M_c & M_d \end{bmatrix}$$

we can write

$$G^T M^T M(I - C^\dagger C) =$$
$$\begin{bmatrix} G_d^T M_d^T M_a^- & G_d^T M_d^T M_a^+ & G_d^T M_d^T M_b \hat{I}_b & G_d^T M_d^T M_c & G_d^T M_d^T M_d \hat{I}_d \\ G_c^T M_c^T M_a^- & G_c^T M_c^T M_a^+ & G_c^T M_c^T M_b \hat{I}_b & G_c^T M_c^T M_c & G_c^T M_c^T M_d \hat{I}_d \end{bmatrix} = 0$$

It is easy to show that $G_c^T M_c^T M_c = 0$ if and only if

$$G_c^T M_c^T = 0 \tag{30}$$

Obviously, M will be singular if M_c is to satisfy (30) which is a contradiction. Therefore, n_c must be zero and therefore no G_c,and M_c appears in SCB form of the system. From (25) and (26),

$$\hat{I}_d = \begin{bmatrix} C_{q1} & 0 & \dots & 0 \\ 0 & \hat{C}_{q2} & \dots & 0 \\ \dots & \dots & \dots & 0 \\ 0 & 0 & 0 & \hat{C}_{qd} \end{bmatrix}$$

where $\hat{C}_{q_i} = I_{qi} - C_{qi}^\dagger C_{qi} = \begin{bmatrix} 0 & 0 \\ 0 & I_{q_i - 1} \end{bmatrix}$. In the special case where $q_i = 1$ for all $i = 1, ..., m_d$, or the order of all infinite zeros equal to one, then $\hat{I}_d = 0$. However, if any $q_i > 1$, it is easy to show that $G_d^T M_d^T M_d \hat{I}_d = 0$ if and only if

$$G_d^T M_d^T = 0 \tag{31}$$

Obviously, again in this case M will be singular if M_d is to satisfy (31) which again is a contradiction. Therefore, $q_i = 1$ for all $i = 1, ..., m_d$ so that $\hat{I}_d = 0$ and $G_d^T M_d^T M_d \hat{I}_d = 0$ is satisfied for any M_d. Finally, the required conditions, namely, $n_c = 0$ and the order of all infinite zeros equal to one are equivalent to $rank(CG) = rank(G)$. Next we show that system (A, G, C) must be minimum phase. Now $G = \begin{bmatrix} 0 \; 0 \; 0 \; I_{n_d} \end{bmatrix}$ and

$$I - C^{\dagger}C = \begin{bmatrix} I_{n_a^-} & 0 & 0 & 0 \\ 0 & I_{n_a^+} & 0 & 0 \\ 0 & 0 & \hat{I}_{n_b} & 0 \\ 0 & 0 & 0 & 0 \end{bmatrix}$$

By writing P as

$$\begin{bmatrix} P_{11} & P_{12} & P_{13} & P_{14} \\ P_{12}^T & P_{22} & P_{23} & P_{24} \\ P_{13}^T & P_{23}^T & P_{33} & P_{34} \\ P_{14}^T & P_{24}^T & P_{34}^T & P_{44} \end{bmatrix}$$

and evaluating (28), it follows that

$$P_{14} = 0 \text{ and } P_{24} = 0$$

On the other hand, it can easily be verified that

$$A - KC = \begin{bmatrix} A_{aa}^- & 0 & \star & \star \\ 0 & A_{aa}^+ & \star & \star \\ 0 & 0 & \star & \star \\ \star & \star & \star & \star \end{bmatrix}$$

for any K, where "$\star$" represents appropriately dimensioned matrices, otherwise non relevant to the proof. Let

$$P_a = \begin{bmatrix} P_{11} & P_{12} \\ P_{12}^T & P_{22} \end{bmatrix}, A_a = \begin{bmatrix} A_{aa}^- & 0 \\ 0 & A_{aa}^+ \end{bmatrix}$$

It can be shown that

$$P(A - KC) = \begin{bmatrix} P_a A_a & \star & \star \\ \star & \star & \star \\ \star & \star & \star \end{bmatrix}$$

According to Lemma 4, P_a must be a Lyapunov matrix of A_a in order that P is definite positive, thus A_a must be a stable matrix. Based on Property 2 stated earlier, invariant zeros of system (A, G, C) are eigenvalues of A_a, which means the system must be minimum phase.

Sufficiency. By assumption, the SCB form of the system is reduced to

$$A = \begin{bmatrix} A_{aa}^- & L_{ab}^- C_b & L_{ad}^- \\ 0 & A_b & L_{bd} \\ E_a^- & E_b & A_d \end{bmatrix}, G = \begin{bmatrix} 0 \\ 0 \\ I_{n_d} \end{bmatrix}, C = \begin{bmatrix} 0 & 0 & I_{n_d} \\ 0 & C_b & 0 \end{bmatrix}$$

Based on the SCB theory [42] $\{A_b, C_b\}$ and $\{A_d, I_{n_d}\}$ form observable pairs. Let K_b and K_d be gains such that $\hat{A}_b = A_b - K_b C_b$ and $\hat{A}_d = A_d - K_d$ are stable. Define the feedback gain K as

$$K = \begin{bmatrix} L_{ad}^- & L_{ab}^- \\ L_{bd} & K_b \\ K_d & 0 \end{bmatrix} \tag{32}$$

then

$$A_0 = A - KC = \begin{bmatrix} A_{aa}^- & 0 & 0 \\ 0 & \hat{A}_b & 0 \\ E_a^- & E_b & \hat{A}_d \end{bmatrix}$$

Let

$$A_{ab} = \begin{bmatrix} A_{aa}^- & 0 \\ 0 & \hat{A}_b \end{bmatrix}, E_{ab} = \begin{bmatrix} E_a^- & E_b \end{bmatrix} \tag{33}$$

Given any positive definite matrix Q_d, compute the Lyapunov matrix P_d of $\hat{A}_d$. Then calculate the spectral radius ρ_d of matrix $(P_d E_{ab})^T P_d E_{ab} Q_d^{-1}$. Define Q as,

$$Q = \begin{bmatrix} \alpha I & P_d E_{ab} \\ (P_d E_{ab})^T & Q_d \end{bmatrix} \tag{34}$$

According to Lemma 3, Q is positive definite for $\alpha > \rho_d$. Let P_{ab} be a solution for $P_{ab} A_{ab} + A_{ab}^T P_{ab} = -\alpha I$. then it can be verified

$$P = \begin{bmatrix} P_{ab} & 0 \\ 0 & P_d \end{bmatrix} \tag{35}$$

is a Lyapunov matrix of A_0 corresponding to the Q which is given by (34) and $G^T P(I - C^\dagger C) = 0$. ■

Lyapunov based Walcott and Zak SMO design algorithm

The proof procedure of Theorem 1 implies an algorithm for the construction of a Walcott and Zak observer that is suitable for numerical implementation. The algorithm is detailed below:

i. If $rank(CG) = rank(G)$ and system (A, G, C) is minimum phase, transform it into SCB form by non-singular state, output, and input transformations $\Gamma_1, \Gamma_2, \Gamma_3$.
ii. Compute K_d and K_b to stabilize $\hat{A}_b = A_b - K_b C_b$ and $\hat{A}_d = A_d - K_d$ respectively. Formulate A_{ab}, E_{ab} as (33).
iii. Compute Lyapunov matrix P_d of $\hat{A}_d$ corresponding to identity matrix, calculate spectral radius ρ_d of matrix $(P_d E_{ab})^T P_d E_{ab}$, then find Lyapunov matrix P_{ab} of A_{ab} corresponding to αI with $\alpha > \rho_d$.

iv. The feedback gain is $K = \Gamma_1 \overline{K} \Gamma_2^{-1}$, where $\overline{K}$ is constructed as in (32). The Lyapunov matrix for $A - KC$ is $P = \Gamma_1^{-T} \overline{P} \Gamma_1^{-1}$, where $\overline{P}$ is given by (35), and W is solved by (27).

Remark 1. It is interesting to note that the existence conditions for unknown inputs observer [33] are exactly the same as those stated in Theorem 1. We shall elaborate on this fact in the next subsection.

2.5 Alternative Walcott-Zak SMO Design

The analysis and the discussion of the last subsection clearly signals that the two required conditions for the design of Walcott and Zak SMO, i.e., (14) and (15) in a round about way impose some structural constraints on the system under consideration. Recently, an explicit equivalent condition for (14) and (15), in terms of original system matrices was derived by Corless and Tu [11]. However, Corless and Tu's work focused on robust state and input estimator design and did not address the connection of their result to sliding mode observer. In the following we shall present a summary of Corless and Tu [11] results as it pertains to the discussion here, in the form of two lemmas. It should be mentioned however that both results have been addressed, albeit using a different and perhaps more general approach than that of [11] in the last two subsections.

Lemma 4. *There exists symmetric positive definite matrices P, and Q and gain matrices K, and W that satisfy (14) and (15) if and only if $rank(CG) = rank(G)$ and the triplet $\{A, G, C\}$ is minimum phase.*

The system satisfying the conditions stated in Lemma 4 has a canonical form, which is described in the following Lemma.

Lemma 5. *For system (2) there exist non-singular transformations Γ_1, and Γ_2 such that*

$$x = \Gamma_1 [x_{ab}^T, x_d^T]^T, y = \Gamma_2 [y_d^T, y_b^T]^T,$$

in which case the transformed system can be written as

$$\begin{aligned} \dot{x}_{ab} &= A_{ab} x_{ab} + L_d x_d + B_{ab} u \\ \dot{x}_d &= E_{ab} x_{ab} + A_d x_d + B_d u + G_d d(x, u, t) \\ y_b &= C_{ab} x_{ab} \\ y_d &= C_d x_d \end{aligned} \tag{36}$$

where G_d is full rank, and C_d is invertible, and the pair (A_{ab}, C_{ab}) is detectable if and only if $rank(CG) = rank(G)$ and the triplet $\{A, G, C\}$ is minimum phase.

Lemma 5 is a direct result of the work on the special coordinate transformation (SCB) theory [42]. Lemmas 4 and 5 imply that Walcott-Zak observer can actually be designed for systems which can be transformed into the form given in (36). Based on the above representation of the system in (36), we shall now suggest a Walcott and Zak type of SMO. It should be noted that the nonlinear term $Gsign(WCe)$ in (36), actually appears in the subsystem which is affected directly by unknown inputs. Letting $K = \begin{bmatrix} K_{ab} & 0 \\ 0 & K_d \end{bmatrix}$, the observer has following form,

$$\begin{aligned} \dot{\hat{x}}_{ab} &= A_{ab}\hat{x}_{ab} + L_d\hat{x}_d + B_{ab}u + K_{ab}(y_b - \hat{y}_b) \\ \dot{\hat{x}}_d &= E_{ab}\hat{x}_{ab} + A_d\hat{x}_d + B_d u + K_d(y_d - \hat{y}_d) + \rho sign(y_d - \hat{y}_d) \end{aligned} \tag{37}$$

Obviously, by making a suitable choice of suitable K_d and ρ, it is possible to make $\hat{x}_d - x_d$ to converge exponentially to zero in spite of the presence of the unknown inputs. So long that $\{A, G, C\}$ is minimum phase, there exists K_{ab} that can make $\hat{x}_{ab} - x_{ab}$ converge to zero. This can be carried out using classical Luenberger observer design. This procedure has considerably simplified the Walcott-Zak SMO design because no constrained Lyapunov equation is involved in the design process.

Remark 2. Note once again that the existence conditions for unknown inputs observers (UIO) [33,23] are exactly the same as those stated in Lemma 4. This is interesting in that it was generally perceived that the sliding mode observers can be designed under less restrictive conditions than UIOs. Considering the the fact that the dynamics of UIO is much simpler than that of Walcott-Zak observer, in addition to the fact that in the case of the UIO the assumption of boundedness of unknown inputs is unnecessary, these results raise certain questions with regards to the applicability and/or advantages of the SMOs. It is also noteworthy to point out that UIO and SMO rely on different operating principles to achieve their robustness to matched uncertainties. Typically when designing an UIO, it is not necessary to estimate the states of x_d subsystem which is subject to unknown inputs, because they are available through the measurement. For an observer based control system design it is unnecessary to estimate those states whose value can be derived directly from outputs. However, for fault diagnosis purposes, the estimation of those measurable and unknown input driving states through sliding mode method may prove to be very useful. Thus it is anticipated that a fault diagnosis scheme based on SMO may have additional capabilities. This issue will be explored in the sequel.

Remark 3. To conclude the discussion of this section, a comparison between the SMO based on the walcott and Zak design versus that of the Utkin's equivalent control principle is in order. The Walcott-Zak based sliding-mode observer (16) design makes $\hat{x} \to x$. Because $d(x, u, t)$ may represent both nonlinearities and unknown inputs due to uncertainties (modeling error or disturbance), it means the Walcott-Zak SMO (16) results in an exact estimation for certain class of nonlinear uncertain systems. If the number of unknown inputs are equal to outputs, it will require the unknown input free subsystem itself to be stable. In this approach no equivalent control information is used. The main disadvantage of the Walcott-Zak observer is that it imposes strong structural constraint on the system and thus is of limited application. On the other hand, the SMOs designed based on the equivalent control principle can work under less stringent existence conditions. However, section 2.1 showed that the drawback with these SMOs comes from the fact that these SMOs can introduce unknown inputs into error dynamics for those unmeasurable states. As a result there exist bounded estimation error for bounded modeling errors when equivalent control based sliding mode observers are employed.

In the next section we propose a novel sliding mode observer design technique by exploiting the structural property of subsystems and by careful use of the equivalent control signal. The proposed design significantly reduces the structural constraint required by the previous design approaches.

3 A Sliding Mode Observer for Linear Uncertain Systems

Consider the SCB form of (2). We propose the following sliding mode observer for each x_{id} subsystem in (23) based on equivalent control method,

$$\begin{aligned}
\dot{\hat{x}}_1^i &= \hat{x}_2^i + L_1^i y_d + B_1^i u + \lambda_1^i \text{sign}(y_i - \hat{x}_1^i) \\
\dot{\hat{x}}_2^i &= \hat{x}_3^i + L_2^i y_d + B_2^i u + \lambda_2^i \text{sign}(\overline{e_2^i}) \\
\cdots & \quad \cdots \\
\dot{\hat{x}}_{q_i-1}^i &= \hat{x}_{q_i}^i + L_{q_i-1}^i y_d + B_{q_i-1}^i u + \lambda_{q_i-1}^i sign(\overline{e_{q_i-1}^i}) \\
\dot{\hat{x}}_{q_i}^i &= E_{ia}\hat{x}_a + E_{ib}\hat{x}_b + E_{ic}\hat{x}_c + \textstyle\sum_{j=1}^{m_d} E_{ij}\hat{x}_{jd} + B_{q_i}^i u + \lambda_{q_i}^i sign(\overline{e_{q_i}^i})
\end{aligned} \tag{38}$$

where $\hat{x}_a, \hat{x}_b, \hat{x}_c$ are estimations for the states x_a, x_b, x_c respectively, coming from sub-observers which are given later. And

$$\overline{e_j^i} = (\lambda_{j-1}^i sign(\overline{e_{j-1}^i}))_{eq} \tag{39}$$

for $j = 2, ..., q_i$, and $e_1 = \hat{x}_1^i - y_i$ can be calculated directly. The equivalent control signal $(v)_{eq}$ for signal v is calculated by a low pass filtering signal v [14]. Further, we do not inject the observation error information before reaching the sliding manifold linked with this information. Moreover, we reach the manifold one by one. More precisely,

$$\overline{e_j^i} = (\lambda_{j-1}^i sign(\overline{e_{j-1}^i}))_{eq} = 0$$

before $\overline{e_{j-1}^i}$ reaches its sliding manifold.

Theorem 3. *For system (23) and observer (38) with any initial state and any bounded unknown input $d_i(x, u, t)$, there exists a choice of*

$$\lambda_j^i, i = 1, ..., m_d, j = 1, ..., q_i$$

such that the state estimation converges in finite time to its real value.

Proof: From (23) and (38), we obtain the following observation error dynamics of $e_j^i = x_j^i - \hat{x}_j^i, j = 1, ..., q_i$,

$$\begin{array}{ll} \dot{e}_1^i & = e_2^i - \lambda_1^i sign(e_1^i) \\ \dot{e}_2^i & = e_3^i - \lambda_2^i sign(\overline{e_2^i}) \\ \cdots & \cdots \\ \dot{e}_{q_i-1}^i & = e_{q_i}^i - \lambda_{q_i-1}^i sign(\overline{e_{q_i-1}^i}) \\ \dot{e}_{q_i}^i & = E_{ia}e_a + E_{ib}e_b + E_{ic}e_c + \sum_{j=1}^{m_d} E_{ij}e_{jd} + d_i - \lambda_{q_i}^i sign(\overline{e_{q_i}^i}) \end{array} \tag{40}$$

where $e_c = x_c - \hat{x}_c, e_a = x_a - \hat{x}_a, e_b = x_b - \hat{x}_b$. Thus, as the known and unknown inputs are bounded, the state does not go to infinity in finite time. Consequently, the observation error state is also bounded. By choosing $\lambda_1^i > |e_2^i|$, e_1^i goes to zero in finite time t_1. Moreover, after t_1, we have

$$\overline{e_2^i} = (\lambda_1^i sign(e_1))_{eq} = e_2^i.$$

Therefore, after t_1, the second error equation becomes

$$\dot{e}_2^i = e_3^i - \lambda_2^i sign(e_2^i).$$

If $\lambda_2^i > |e_3^i|$, e_2 goes to zero in finite time $t_2 > t_1$. Therefore after $t > t_2$,

$$\overline{e_3^i} = (\lambda_2^i sign(e_2))_{eq} = e_3^i,$$

$$\dot{e}_3^i = e_4^i - \lambda_3^i sign(e_3^i).$$

We run the procedure up to step q_i, thus after t_{q_i-1}, we have

$$\dot{e}_{q_i}^i = E_{ia}e_a + E_{ib}e_b + E_{ic}e_c + \sum_{j=1}^{m_d} E_{ij}e_{jd} + d_i - \lambda_{q_i}^i sign(e_{q_i}^i).$$

Let

$$\lambda_{q_i}^i > |E_{ia}e_a + E_{ib}e_b + E_{ic}e_c + \sum_{j=1}^{m_d} E_{ij}e_{jd} + d_i|, \tag{41}$$

e_{q_i} converges to zero in finite time $t_{q_i} > t_{q_i-1}$. ∎

Now, going back to provide the estimation of x_a, x_b and x_c in (38). A classical Luenberger observer is applied to subsystem x_b,

$$\dot{x}_b = A_b x_b + L_{bd} y_d + B_b u, \qquad y_b = C_b x_b$$

because it is unknown input free, and (A_b, C_b) forms an observable pair. Of course, the Utkin SMO can be used as well. The subsystem x_a can be further decomposed into two subsystems,

$$\begin{aligned}\dot{x}_a^- &= A_a^- x_a^- + L_a^- \overline{y} + B_a^- u \\ \dot{x}_a^+ &= A_a^+ x_a^+ + L_a^+ \overline{y} + B_a^+ u\end{aligned}$$

A_a^- is stable, and therefore a Luenberger observer for subsystem x_a^- can be designed. However, because A_a^+ is unstable, some other way has to be found in order to estimate x_a^+.

Let us now discuss the possibility of estimating x_a^+ together with the estimation of x_c. After $e_{q_i}^i$ reaches sliding mode, we have the following equivalent control signal:

$$(\lambda_{q_i}^i sign(e_{q_i}^i))_{eq} = E_{ia}^- e_a^- + E_{ia}^+ e_a^+ + E_{ib} e_b + E_{ic} e_c + \sum_{j=1}^{m_d} E_{ij} e_{jd} + d_i \qquad i = 1, ..., m_d$$

and after all states of x_a^-, x_b and x_d have been estimated, it will equal to

$$(\Lambda sign(e_q))_{eq} = E_{dc} e_c + E_{da}^+ e_a^+ + d_d \tag{42}$$

where all m_d equivalent control signals are written together as a vector. Considering the above equivalent control signal as the output of the x_c, x_a^+ subsystem, this subsystem can be rewritten as

$$\begin{aligned}\dot{x}_{ac} &= A_{ac} x_{ac} + L_{ac} \overline{y} + B_{ac} u + G_{ac} d \\ y_{ac} &= E_{ac} x_{ac} - E_{ac} \hat{x}_{ac} + G_o d\end{aligned} \tag{43}$$

where

$$x_{ac} = \begin{bmatrix} x_a^+ \\ x_c \end{bmatrix}, A_{ac} = \begin{bmatrix} A_a^+ & 0 \\ G_c E_{ca}^+ & A_c \end{bmatrix}, G_{ac} = \begin{bmatrix} 0 & 0 \\ G_c & 0 \end{bmatrix},$$

$$E_{ac} = \begin{bmatrix} E_{da}^+ & E_{dc} \end{bmatrix}, G_o = \begin{bmatrix} 0 & I \end{bmatrix},$$

An interesting fact is that (A_{ac}, E_{ac}) is detectable if (A, C) is detectable. Unfortunately, the system (43) has unknown inputs, and no UIO for this system exist according to the following lemma.

Lemma 6. *[28] A UIO for system (43) exists only if*

$$rank \begin{bmatrix} G_o & E_{ac} G_{ac} \\ 0 & G_o \end{bmatrix} = rank G_o + rank \begin{bmatrix} G_{ac} \\ G_o \end{bmatrix} \tag{44}$$

It is easy to show that condition (44) will never be satisfied due to the particular form of G_s, G_o.

As a last resort, we may use H_2 or H_∞ optimal observer design techniques proposed in [43] to make the estimation error as small as possible. On the other hand, [49] has shown that exact or approximate estimation using sliding mode observer is impossible if all the measurements of a system are disturbed by the unknown inputs. Assuming the disturbance on the measurement is bounded by some constant v_0, the estimation error will not be bigger than v_0.

Finally, although in general we can not design observers to make e_a^+ and e_c become zero, as long as these errors are bounded, their value will not affect the convergence of the sliding mode observer (38) with a gain satisfying (41).

At this point, based on the above analysis, we shall summarize the new sliding mode (functional) observer design procedure for a linear uncertain system as follows.

Sliding mode and sliding mode functional observer design algorithms

The steps of the following algorithm can render either a state or when the state observer design is not possible, a functional sliding mode observer:

i. Transform system (2) it into its SCB form (19)-(20) by non-singular transformations $\Gamma_1, \Gamma_2, \Gamma_3$.
ii. Estimate x_d using the sliding mode observer (38), and estimate x_a^-, x_b using a regular Luenberger observer. The measurement variable for the transformed system y_d, y_b is derived from the original output by $[y_d, y_b]' = \Gamma_2^{-1} y$, and the known input distribution matrix is transformed by $\Gamma_1^{-1} B$.
iii. If $\{A, G, C\}$ is left invertible (or equivalently $n_c = 0$) and is minimum phase (or $n_a^+ = 0$), the original state can be derived by

$$x = \Gamma_1 [x_a^- \quad x_b \quad x_d]^T.$$

iv. If $n_c \neq 0$ or $n_a^+ \neq 0$, any linear function of the states, $T_{k\times n} x$ can be estimated, where T must satisfy the following condition:

$$T\Gamma_1 = [T_a^- \quad 0 \quad T_b \quad 0 \quad T_d]_{k\times n} = \hat{T}_{k\times n}$$

where T_a^-, T_b, T_d are any matrices of dimension $k \times n_a^-, k \times n_b$ and $k \times n_d$ respectively. Obviously, the maximum rank of T is $n_a^- + n_b + n_d$.

Remark 4. The Walcott-Zak SMO requires $rank(CG) = rank(G)$, which immediately implies that $n_c = 0$ and $q_i = 1, i = 1, ..., m_d$, namely, the number of infinite zeros of order i is one. The restriction on the system $\{A, G, C\}$ infinite zero structure is removed in our algorithm. Unfortunately, the requirement of $n_c = 0$ is still necessary for estimating all states, and it implies $rank(C) \geq rank(G)$.

Remark 5. In [60], linear state function Tx is estimated using unknown input functional observer (UIFO), and the maximum rank of T is $n_a^- + n_b$. The maximum rank of T has been increased significantly using our new SMFO design.

Checking the equivalent control signal (42), we have following corollary about unknown inputs estimation.

Corollary 1. *If the system $\{A, G, C\}$ is left invertible (i.e. $n_c = 0$) and all unstable transmission zeros are unobservable modes (i.e. in SCB form, $E_{da}^{+} = 0$), all unknown inputs can be estimated exactly using the proposed SMFO. If system $\{A, G, C\}$ is not left invertible (i.e. $n_c > 0$) but all eigenvalues of corresponding n_c subsystem are unobservable modes (i.e. $E_{dc} = 0$), and all unstable transmission zeros are unobservable modes (i.e. $E_{da}^{+} = 0$), at least m_d unknown inputs can be estimated exactly using the proposed SMFO.*

Remark 6. Compared with input estimator based on UIFO, which was discussed in [11,61], our new SMFO has better capability in estimating the unknown inputs as well.

4 Nonlinear SMO Design Extension

Consider multivariable nonlinear systems described in state space form by equations of the following form

$$\begin{aligned} \dot{x} &= f(x) + B(x,u) + \textstyle\sum_{i=1}^{m} g_i(x) d_i(x,u,t) \\ y_1 &= h_1(x) \\ \dots &\quad \dots \\ y_p &= h_p(x) \end{aligned} \tag{45}$$

in which $x \in M$, a C^∞ connected manifold of dimension n, $f(x), B(x,u)$, and $G(x) = [g_1(x), ..., g_m(x)]$ are smooth vector fields on M, and $h_i(x), i = 1, ..., p$ are smooth functions from M to R. In what follows, local coordinates are generally used. When global properties are considered, notions are simplified by assuming that M accepts a global coordinate system.

Assumption 1 We assume that $p \geq m$, and the first m outputs have a vector relative degree $\{q_1, ..., q_m\}$ corresponding to $G(x)$ at each point $x_0 \in M$. Stated differently, this means

$$L_{g_j} L_f^k h_i(x) = 0$$

for all $j = 1, ..., m$, for all $k < q_i - 1$, for all $i = 1, ..., m$, and for all x in M. Further, the $m \times m$ matrix

$$A(x) = \begin{pmatrix} L_{g_1} L_f^{q_1-1} h_1(x) & \dots & L_{g_m} L_f^{q_1-1} h_1(x) \\ L_{g_1} L_f^{q_2-1} h_2(x) & \dots & L_{g_m} L_f^{q_2-1} h_2(x) \\ \dots & \dots & \dots \\ L_{g_1} L_f^{q_m-1} h_m(x) & \dots & L_{g_m} L_f^{q_m-1} h_m(x) \end{pmatrix}$$

is nonsingular at each point $x_0 \in M$.

Assumption 2 The distribution spanned by the vector fields $g_1(x)$,.., $g_m(x)$ is involutive.

Lemma 7. *Given the system (45), if assumptions 1-2 are valid, then*

$$q_1 + ... + q_m \leq n$$

For $i = 1, ..., m$ set,

$$\begin{aligned} \phi_1^i &= h_i(x) \\ \phi_2^i &= L_f h_i(x) \\ ... &\quad ... \\ \phi_{q_i}^i &= L_f^{q_i - 1} h_i(x) \end{aligned}$$

if $q = q_1 + ... + q_m$ is strictly less than n, it is always possible to find $n - q$ more functions $\phi_{q+1}(x), ..., \phi_n(x)$ such that the mapping

$$\Phi(x) = col(\phi_1^1(x), ..., \phi_{q_1}^1(x), ..., \phi_{q_m}^m(x), \phi_{q+1}(x), ..., \phi_n(x))$$

has a Jacobian matrix which is nonlinear at each $x_0 \in M$ and therefore qualifies as a local coordinate transformation in M. Moreover, based on Assumption 2, it is always possible to choose $\phi_{q+1}(x), ..., \phi_n(x)$ in such a way that

$$Lg_j \phi_i(x) = 0 \tag{46}$$

for all $i = q + 1, ..., n, j = 1, ..., m$, and all x in M.

Set

$$x_d^i = \begin{pmatrix} x_1^i \\ x_2^i \\ ... \\ x_{q_i}^i \end{pmatrix} = \begin{pmatrix} \phi_1^i(x) \\ \phi_2^i(x) \\ ... \\ \phi_{q_i}^i(x) \end{pmatrix}$$

for $i = 1, ..., m$, and $x_d = ((x_d^1)^T, ..., (x_d^m)^T)$,

$$x_o = \begin{pmatrix} x_o^1 \\ x_o^2 \\ ... \\ x_o^{n-q} \end{pmatrix} = \begin{pmatrix} \phi_{q+1}(x) \\ \phi_{q+2}(x) \\ ... \\ \phi_n(x) \end{pmatrix}$$

the equations under new coordinates can be written as

$$\begin{aligned} \dot{x}_1^i &= x_2^i + b_1^i(x_d, x_o, u) \\ ... &\quad ... \\ \dot{x}_{q_i-1}^i &= x_{q_i}^i + b_{q_i-1}^i(x_d, x_o, u) \\ \dot{x}_{q_i}^i &= a_i(x_d, x_o) + b_{q_i}^i(x_d, x_o, u) + \textstyle\sum_{j=1}^m c_{ij}(x_d, x_o) d_j \\ y_i &= x_1^i \end{aligned} \tag{47}$$

for $i = 1, ..., m$, and

$$\begin{aligned} \dot{x}_o &= q(x_d, x_o) + p(x_d, x_o, u) \\ y_{m+1} &= h_{m+1}(x_d, x_o) \\ ... &\quad ... \\ y_p &= h_p(x_d, x_o) \end{aligned} \tag{48}$$

where

$$a_i(x_d, x_o) = L_f^{q_i} h_i(\Phi^{-1}(x_d, x_o)), c_{ij} = L_{g_j} L_f^{q_i - 1} h_i(\Phi^{-1}(x_d, x_o))$$

and

$$b_k^i(x_d, x_o, u) = \frac{\partial(L_f^{k-2} h_i)}{\partial x} B(\Phi^{-1}(x_d, x_o), u)$$

The above lemma is a trivial extension of Proposition 5.1.2 in [29]. The nonlinear transformation $\Phi(x)$ decomposes the system (45) into two subsystems. Next, we shall discuss the observer design for system (47)-(48).

Before one can proceed with the observer design, the observability of nonlinear subsystem (47)-(48) needs to be studied. It is well known that inputs which make a nonlinear system unobservable are called "bad inputs" or "not universal inputs". For known input signals, one can avoid applying those "bad inputs". However, the unknown inputs which may be the result of a fault or certain other disturbances are beyond ones control. Therefore, observability for all unknown inputs is in general necessary in order to design a nonlinear unknown input observer, unless it can be guaranteed *a priori* that the unknown inputs do not belong to the class of bad inputs. Based on the work in [22], we have following lemma that shows the conditions such that the observability of x_d subsystem is independent of unknown inputs.

Lemma 8. *Let* $\overline{x_d^i} = \{x_d^1, ..., x_d^{i-1}, x_d^{i+1}, ..., x_d^m\}$. *Assume each* x_d^i *subsystem in (47) has its input term in the following form*

$$\begin{array}{ll} b_1^i(x_d, x_o, u) & = b_1^i(x_1^i, x_2^i, \overline{x_d^i}, x_o, u) \\ b_2^i(x_d, x_o, u) & = b_2^i(x_1^i, x_2^i, x_3^i, \overline{x_d^i}, x_o, u) \\ \dots & \dots \\ b_{q_i-1}^i(x_d, x_o, u) & = b_{q_i-1}^i(x_d, x_o, u) \\ b_{q_i}^i(x_d, x_o, u) & = b_{q_i}^i(x_d, x_o, u) \end{array} \tag{49}$$

and the functions

$$1 + \frac{\partial b_j^i}{\partial x_{j+1}^i} \neq 0, j = 1, ..., q_i - 1$$

and state $x_o, \overline{x_d^i}$ *is considered as inputs for* x_d^i *subsystem. Then* x_d^i *subsystem is uniformly observable for all inputs* $u, d, \overline{x_d^i}$ *and* x_o.

4.1 SMO for Subsystem with Unknown Inputs

We shall first build an unknown input independent observer for x_d subsystem.

Theorem 4. *If each x_d^i subsystem in (47) has its input term in the following form*

$$\begin{array}{ll} b_1^i(x_d, x_o, u) & = b_1^i(y, u) \\ b_2^i(x_d, x_o, u) & = b_2^i(x_2^i, y, u) \\ \dots & \dots \\ b_{q_i-1}^i(x_d, x_o, u) & = b_{q_i-1}^i(x_2^i, \dots, x_{q_i-1}^i, y, u) \\ b_{q_i}^i(x_d, x_o, u) & = b_{q_i}^i(x_d, x_o, u) \end{array} \tag{50}$$

there exists a choice of $\lambda_j^i, i = 1, ..., m, j = 1, ..., q_i$ such that following sliding mode observer can be built to estimate states,

$$\begin{array}{ll} \dot{\hat{x}}_1^i & = \hat{x}_2^i + b_1^i(y, u) + \lambda_1^i sign(y_i - \hat{x}_1^i) \\ \dot{\hat{x}}_2^i & = \hat{x}_3^i + b_2^i(\hat{x}_2^i, y, u) + \lambda_2^i sign(\overline{e_2^i}) \\ \dots & \dots \\ \dot{\hat{x}}_{q_i-1}^i & = \hat{x}_{q_i}^i + b_{q_i-1}^i(\hat{x}_2^i, \dots, \hat{x}_{q_i-1}^i, y, u) + \lambda_{q_i-1}^i sign(\overline{e_{q_i-1}^i}) \\ \dot{\hat{x}}_{q_i}^i & = a_i(\hat{x}_d, \hat{x}_o) + b_{q_i}^i(\hat{x}_d, \hat{x}_o, u) + \lambda_{q_i}^i sign(\overline{e_{q_i}^i}) \end{array} \tag{51}$$

where $\overline{e_j^i}, j = 1, ..., q_i$ are given by

$$\overline{e_j^i} = (\lambda_{j-1}^i sign(\overline{e_{j-1}^i}))_{eq} \tag{52}$$

for $j = 2, ..., q_i$, and $\overline{e_1^i} = e_1^i = y_i - \hat{x}_1^i$ can be obtained directly.

Proof: Follows along the same lines as the linear case ([63]). ∎

Once again, the equivalent control signal $(v)_{eq}$ for signal v is calculated by low pass filtering the signal v with anti-peaking structure [14]. This anti-peaking structure is based on the principle that the observation error information is not used before reaching the sliding manifold linked with this information. Moreover, we reach the manifold one by one in a sort of a recursive fashion. As a result of this, obtain a sub-dynamics of dimension one is encountered and consequently the peaking phenomena is avoided. More precisely,

$$\overline{e_j^i} = (\lambda_{j-1}^i sign(\overline{e_{j-1}^i}))_{eq} = 0$$

before $\overline{e_{j-1}^i}$ reaching its sliding manifold.

Remark 7. Note that the transformation in Lemma 7 decomposed the nonlinear system (45) into two interconnected subsystem with states x_d and x_o. Of these two subsystems, the unknown inputs only appear in one, namely the x_d. The observer proposed in the above discussion will provide state estimates for this subsystem. On the other hand, the x_o subsystem is free of the unknown inputs, and any applicable existing observer can be used to estimate its state $\hat{x}_o$. This estimate is needed in observer (51), for estimating the state of x_d. On the other hand, if x_o subsystem is unobservable, or if it is too complex to build an observer for it, $\hat{x}_o$ will not approximate real value of x_o. However, as long as both $\hat{x}_o$ and x_o are bounded, observer (51) will always converge as long as gain $\lambda^i_{q_i}$ is selected large enough.

It is therefore possible to let $\hat{x}_o = 0$ and still estimate part of the state. The reason for this desirable property is that x_o is only confined to the last equation. In such a case, we are essentially building a functional sliding mode observer. It is well known, at least in the linear realm, that it is generally possible to design functional observers under less restrictive conditions than full state observers. Such observers can be useful for control purposes and more so for fault diagnosis applications. In fault diagnosis applications observers are essentially used as residual generators. The estimation error of the observer is used as a residual that is used to draw conclusions about the health of the system. In such applications, a full estimate of the actual state may not be necessary and a functional observer may accomplish the fault detection task under less restrictive existence conditions [62].

Remark 8. If $q_1 + ... + q_m = n$, then the x_o subsystem will actually disappear, and all states can be estimated.

Remark 9. It should be noted that if $q_i = 1, i = 1, ..., m$, the above theorem does not put any special constraint on the input term. $q_i = 1, i = 1, ..., m$ means all states of subsystem x_d are measurable. An observer for such a system may be unnecessary for controller synthesis, but will be useful in fault detection and isolation (FDI) applications.

To conclude this section, although sliding mode concept provides a nice framework for observer design of a general class of nonlinear uncertain systems, due to the inherent property of sliding mode, there exist two basic drawbacks for practical applications. First, although the bound of unknown input is known, the estimation error bound is not known *a priori*. This makes the selection of the gain difficult. If the gain is too large, observer will exhibit excessive chattering before reaching sliding mode. If the gain is too small, observer may never be able to reach sliding mode and converge to real state value. Secondly, it is difficult to choose a suitable time to use the equivalent control signal. The equivalent control signal should be used only if its corresponding estimation error is near zero, or in sliding mode. However, except for those states which are measured, we do not know if a state estimation is in sliding mode or not. If the equivalent control is used too early, peaking phenomena is unavoidable. If the equivalent control is used too late, a correct estimation in time cannot be achieved which is unacceptable for high quality control.

5 A Simple Sliding Mode Output Observer for Fault Diagnosis

Most observer-based fault diagnosis schemes generate residuals by comparing the measurement and its corresponding estimate provided by observers.

Because most observers will produce wrong estimates when faults occur, a nonzero residual would raise an alarm. In this section it is shown that sliding mode output observers can produce the estimates of outputs for general nonlinear systems, and can be useful for detecting faults of large magnitude. However, it will be shown that the SMOO framework is not useful for the detection of incipient faults in many practical applications where system model mismatch is unavoidable and is significant. An incipient fault slowly develops from zero to a certain significant value, it will neither produce a sudden peak in the residual, nor will it push the system out of sliding mode, at least when its value is small in its early stage. Many previous studies on SMO based fault diagnosis neglected this important limitation.

Consider a multivariable nonlinear system described in state space form by equations of the following kind:

$$\begin{aligned} \dot{x} &= A(x, u, d, f_a(t)) \\ y &= H(x) + f_s(t) \end{aligned} \tag{53}$$

where $x \in M$, a C^∞ connected manifold of dimension n, and $H(x) = [h_1(x), ..., h_p(x)]$ are smooth vector fields on M, u and d represent the control input and the signal representing the system uncertainties respectively. It is assumed that all x, u and d are bounded. The term $f_a(t)$ represents actuator and component faults, while $f_s(t) = [f_s^1(t), ..., f_s^p(t)]$ represent the sensor faults, which can be any function of time.

A general sliding mode output observer (SMOO) for the above system is of the following form,

$$\dot{z} = L(y, u) + \Lambda sign(y - z) \tag{54}$$

where z is an estimate of $H(x)$, $L(y, u)$ is a function of y and u, Λ is a diagonal gain matrix with elements $\lambda_i, i = 1, ..., p$. The design of $L(y, u)$ is discussed later.

Based on the above, the dynamics of the estimation error $e = H(x) - z$ can be obtained

$$\begin{aligned} \dot{e} = & \frac{\partial H(x)}{\partial x} A(x, u, d, f_a(t)) \\ & - L(H(x) + f_s(t), u) - \Lambda sign(H(x) + f_s(t) - z) \end{aligned} \tag{55}$$

Note that e cannot be measured, and only $r = y - z = H(x) + f_s(t) - z$ is measurable. The quantity r is the sliding surface and can be considered as a residual signal. Assume $e_i = h_i(x) - z_i$ and defining

$$\frac{\partial H(x)}{\partial x} A(x, u, d, f_a(t)) = \begin{bmatrix} m_1(x, u, d, f_a(t)) \\ m_2(x, u, d, f_a(t)) \\ \cdots \\ m_p(x, u, d, f_a(t)) \end{bmatrix} \quad \text{and } L(y, u) = \begin{bmatrix} l_1(y, u) \\ l_2(y, u) \\ \cdots \\ l_p(y, u) \end{bmatrix}.$$

the dynamics of e_i can be expressed as

$$\dot{e}_i = m_i(x, u, d, f_a(t)) - l_i(H(x) + f_s(t), u) - \lambda_i sign(e_i(x) + f_s^i(t)).$$

Assuming that no sensor or actuator faults are present, i.e. $f_a(t) = f_s(t) = 0$, and if $|m_i(x,u,d,0) - l_i(H(x),u)| \leq \lambda_i$, we have

$$\frac{d}{dt}e_i^2 = 2e_i\dot{e}_i = 2e_i(m_i(x,u,d,0) - l_i(H(x),u) - \lambda_i sign(e_i)).$$

Therefore,

$$\begin{aligned}&\text{if } e_i > 0,\ \tfrac{d}{dt}e_i^2 = 2e_i(m_i(x,u,d,0) - l_i(H(x),u) - \lambda_i) < 0;\\ &\text{if } e_i < 0,\ \tfrac{d}{dt}e_i^2 = 2e_i(m_i(x,u,d,0) - l_i(H(x),u) + \lambda_i) < 0.\end{aligned}$$

Thus $r_i = e_i$ exponentially decreases to zero according to the Lyapunov principle. Next we shall investigate what happens when sensor or actuator faults occur. It is necessary to determine how a fault interacts with the sliding surfaces and how the sliding performance of the observer is affected.

Case 1: Effect of sensor faults

If sensor i becomes faulty at time t_i, namely $f_s^i(t) \neq 0$ after $t > t_i$, the residual $r_i = e_i + f_s^i(t)$, then

$$\dot{r}_i = m_i(x,u,d,0) - l_i(H(x) + f_s(t), u) + \dot{f}_s^i(t) - \lambda_i sign(r_i). \tag{56}$$

In this case the sensor fault may produce one of two results:

i. If $|m_i(x,u,d,0) - l_i(H(x) + f_s(t),u) + \dot{f}_s^i(t)| \leq \lambda_i$, the sliding behavior can then only occur on the surface

$$r_i = h_i(x) + f_s^i(t) - z_i = 0. \tag{57}$$

In other words, the estimation error $h_i(x) - z_i$ is nonzero. However, the estimation error cannot be measured, and the measured residual signal will still be zero. Fortunately, if the fault is an abrupt event, or $f_s^i(t)$ jumps from zero to a significantly large value at time t_i (see Figure 2a), a corresponding abrupt change will be observed in r_i. The reason for such a phenomena is that the residual will change from zero to $f_s^i(t_i)$, and then decrease to zero quickly. Unfortunately, not all abrupt changes of $f_s^i(t)$ may represent faults that need to be detected. Consider the shape of f_s^i shown in Figure 2b, which represents a sensor jitter. This sensor jitter will produce almost the same residual shape as an abrupt sensor fault. In such a case, it is difficult to distinguish between the real abrupt sensor fault and sensor jitter.

ii. If

$$|m_i(x,u,d,0) - l_i(H(x) + f_s(t),u) + \dot{f}_s^i(t)| > \lambda_i \tag{58}$$

persistently holds, the observer will be disturbed from its surface and sliding will cease. In this case, the ith residual element will become nonzero persistently, and will alarm the occurrence of a sensor fault.

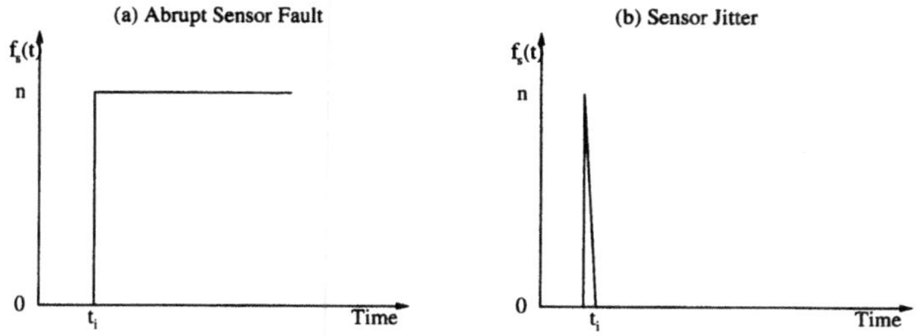

Fig. 2. Two possible sensor failure modes

For multiple sensor fault identification, we need to design $l_i(y,u) = l_i(y_i,u)$ so that r_i is only sensitive to $f_s^i(t)$.

Case 2: Effect of actuator faults

If actuators fail, namely if $f_a(t) \neq 0$ when $t > t_a$, we may also observe two effects on the residual:

i. If $|m_i(x,u,d,f_a(t)) - l_i(H(x),u)| \leq \lambda_i$ for all $i = 1, ..., k$, the observer, will keep on the sliding surface and faults can not be detected, i.e.,

$$r_i = e_i = 0, i = 1, ..., k.$$

ii. If

$$|m_i(x,u,d,f_a(t)) - l_i(H(x),u)| > \lambda_i \tag{59}$$

is true persistently for some $i \in \{1,2,...k\}$, the residual element, r_i, will become nonzero which would indicate an onset of a failure.

In practice, due to ever present measurement noise, saturation function could be used to replace the sign function. When there are no faults present, the residual is not exactly zero, but less than a small threshold. Once its value passes the threshold, an alarm is registered.

Amongst the past FDI studies, [52] only considered the abrupt sensor fault detection, and [25] simply assumed that the faults were persistent and had moved outside the interval of robustness. While these studies are based on the linearized system model, the above discussion has conceptually shown that fault diagnosis using the sliding mode concept can actually be used for general nonlinear uncertain systems, as long as

i. The upper bound of model mismatch $|m_i(x,u,d,0) - l_i(H(x),u)|$ is known *a priori*, such that a suitable gain λ_i can be selected;
ii. The faults result in a system model mismatch out of the boundary, namely condition (58) or (59) is satisfied;

Therefore, in order to improve the capability of SMOO based fault diagnosis scheme, we must minimize the effect of the model mismatch through the design of $L(y,u)$. Recall the error dynamics of SMOO given by (55), we know that $L(y,u)$ should be designed such that

$$\|\frac{\partial H(x)}{\partial x}A(x,u,d,0) - L(H(x),u)\|$$

is minimized for all possible value of state x and input u. The solution of $L(y,u)$ will be strongly dependent on the specific form of the nonlinear equation (53). Consider a simple case of linear uncertain systems described by the following equation:

$$\begin{aligned} \dot{x} &= Ax + Bu + Gd + F_a f_a \\ y &= Cx + F_s f_s. \end{aligned} \tag{60}$$

With the general form of SMOO (54), the estimation error equation (55) is simplified as

$$\begin{aligned} \dot{e} = {} & CAx + CBu + CGd + CF_a f_a - L(Cx + F_s f_s(t), u) - \\ & \Lambda sign(Cx + F_s f_s - z). \end{aligned}$$

For the linear system (60), $L(y,u)$ is designed as $L(y,u) = Ky + CBu$. The problem of minimizing the model mismatch is formulated as finding the gain K, such that $\|CA - KC\|$ is minimized. Obviously, the residual is totally independent of the unknown input d and the unknown states x, if and only if $CG = 0$ and there exists K satisfying $CA - KC = 0$, namely

$$\dot{e} = CF_a f_a - KF_s f_s(t) - \Lambda sign(e + F_s f_s).$$

However this is impossible in most cases.

6 Incipient Fault Diagnosis Using SMFO for Linear Uncertain Systems

In order to detect incipient faults, we have to exploit the structural properties of the system. Therefore, we use special coordinate basis (SCB) to transform the linear uncertain system (60). The SCB transformation is applied to matrices (A, G, C) of the system (60). It is easy to show that the fault distribution matrices F_a and F_s will be transformed as

$$\overline{F_a} = \Gamma_1^{-1} F_a = \left[F_{aa}^T \; F_{ab}^T \; F_{ac}^T \; F_{ad}^T \right]^T, \tag{61}$$

and

$$\overline{F_s} = \Gamma_2^{-1} F_s = \begin{bmatrix} F_{sd} \\ F_{sb} \end{bmatrix}. \tag{62}$$

Therefore, the SCB form of the system (60) will be

$$\begin{aligned}\dot{x}_a &= A_a x_a + L_{ad}C_d x_d + L_{ab}C_b x_b + B_a u + F_{aa} f_a,\\ \dot{x}_b &= A_b x_b + L_{bd}C_d x_d + B_b u + F_{ab} f_a, \qquad y_b = C_b x_b + F_{sb} f_s,\\ \dot{x}_c &= A_c x_c + L_{cd}C_d x_d + L_{cb}C_b x_b + G_c E_{ca} x_a + G_c d_c + B_c u + F_{ac} f_a\end{aligned} \tag{63}$$

and $x_d = \left[x_{1d}\; x_{2d}\; ...\; x_{m_d d}\right]^T$. Each $x_{id}, i = 1, ..., , m_d$ is further composed of q_i states, or

$$x_{id} = \left[x_1^i\; x_2^i\; ...\; x_{q_i}^i\right]^T;$$

so $F_{ad} = [F_{d1}^1, ..., F_{dq_1}^1, ..., F_{d1}^{m_d}, ..., F_{dq_{m_d}}^{m_d}]^T$, and the equations of x_{id} are

$$\begin{aligned}\dot{x}_1^i &= x_2^i + L_1^i C_d x_d + B_1^i u + F_{d1}^i f_a\\ \dot{x}_2^i &= x_3^i + L_2^i C_d x_d + B_2^i u + F_{d2}^i f_a\\ &\cdots\\ \dot{x}_{q_i-1}^i &= x_{q_i}^i + L_{q_i-1}^i C_d x_d + B_{q_i-1}^i u + F_{d(q_i-1)}^i f_a\\ \dot{x}_{q_i}^i &= E_{ia} x_a + E_{ib} x_b + E_{ic} x_c + \textstyle\sum_{j=1}^{m_d} E_{ij} x_{jd} + B_{q_i}^i u + F_{dq_i}^i f_a + d_i\\ y_{id} &= x_1^i + f_{sd}^i\end{aligned} \tag{64}$$

where f_{sd}^i is the ith element of $F_{sd} f_s$. Note that $y_d = C_d x_d$ if no sensor faults occur.

6.1 Residual Generation Using Utkin SMO for Unknown Input Free Subsystem

Here we discuss if a SMO can be used to generate the residuals if observable and unknown input free subsystem, x_b, does exist. It is well known that an Utkin SMO can be built for a linear system without unknown input. In order to design an Utkin SMO, we further transform the x_b subsystem into following form:

$$\begin{aligned}\dot{x}_{b1} &= A_{11} x_{b1} + A_{12} x_{b2} + L_{bd1} C_d x_d + B_{b1} u + F_{ab1} f_a\\ \dot{x}_{b2} &= A_{21} x_{b1} + A_{22} x_{b2} - L_{bd2} C_d x_d + B_{b2} u + F_{ab2} f_a\\ y_b &= x_{b2} + F_{sb} f_s.\end{aligned} \tag{65}$$

The corresponding Utkin SMO is

$$\begin{aligned}\dot{\hat{x}}_{b1} &= A_{12} y_b + A_{11}\hat{x}_{b1} + L_{bd1} y_d + B_{b1} u + \Lambda_2(\Lambda_1 sign(y_b - \hat{x}_{b2}))_{eq}\\ \dot{\hat{x}}_{b2} &= A_{22}\hat{x}_{b2} + A_{21}\hat{x}_{b1} + L_{bd2} y_d + B_{b2} u + \Lambda_1 sign(y_b - \hat{x}_{b2})\end{aligned} \tag{66}$$

where Λ_1 is a nonsingular diagonal matrix and its diagonal elements must be large enough to compensate initial estimation errors. $(\Lambda_1 sign(y_b - \hat{x}_{b2}))_{eq}$ is the equivalent control of $\Lambda_1 sign(y_b - \hat{x}_{b2})$, Λ_2 is a matrix selected to make $A_{11} - \Lambda_2 A_{21}$ stable.

The estimation error equations for the observer (66) are

$$\begin{aligned}\dot{e}_{b1} &= -A_{12} F_{sb} f_s + A_{11} e_{b1} - L_{bd1} F_{sd} f_s + F_{ab1} f_a -\\ &\quad \Lambda_2(\Lambda_1 sign(e_{b2} + F_{sb} f_s))_{eq}\\ \dot{e}_{b2} &= A_{22} e_{b2} + A_{21} e_{b1} - L_{bd2} F_{sd} f_s + F_{ab2} f_a - \Lambda_1 sign(e_{b2} + F_{sb} f_s)\end{aligned} \tag{67}$$

where $e_{b2} = x_{b2} - \hat{x}_{b2}, e_{b1} = x_{b1} - \hat{x}_{b1}$.

Now, we set the residual as $r = y_b - \hat{x}_{b2} = e_{b2} + F_{sb}f_s$. Its dynamic equation is

$$\dot{r} = A_{22}r - A_{22}F_{sb}f_s + A_{21}e_{b1} - L_{bd2}F_{sd}f_s - F_{sb}\dot{f}_s + F_{ab2}f_a - \Lambda_1 sign(r).$$

If faults are incipient, it is expected that r will keep on the sliding plane, thus

$$\begin{aligned} r_{eq} &= (\Lambda_1 sign(r))_{eq} \\ &= -A_{22}F_{sb}f_s + A_{21}e_{b1} - L_{bd2}F_{sd}f_s - F_{sb}\dot{f}_s + F_{ab2}f_a \\ &= A_{21}e_{b1} + f_{w2}, \end{aligned} \tag{68}$$

then

$$\begin{aligned} \dot{e}_{b1} &= (A_{11} - \Lambda_2 A_{21})e_{b1} + A_{12}F_{sb}f_s - L_{bd1}F_{sd}f_s + F_{ab1}f_a - \Lambda_2 f_{w2} \\ &= (A_{11} - \Lambda_2 A_{21})e_{b1} + f_{w1} - \Lambda_2 f_{w2}. \end{aligned}$$

Finally, we have the following transfer function of r_{eq},

$$r_{eq}(s) = A_{21}(sI - A_{11} + \Lambda_2 A_{21})^{-1}(f_{w1}(s) - \Lambda_2 f_{w2}(s)) + f_{w2}(s). \tag{69}$$

r_{eq} will be the residual that is used for fault detection. r_{eq} is a function of the fault signal and is independent of unknown inputs or unknown state variable. In practice, we can easily check if r_{eq} will be nonzero for certain faults using equation (69).

Remark 10. If C_b happens to be of full rank, the above Utkin SMO reduces into the simple SMOO discussed in Section 5. Because e_{b1} term actually disappears under this condition, r_{eq} in equation (68) is simplified as

$$r_{eq} = -A_b F_{sb} f_s - L_{bd} F_{sd} f_s - F_{sb}\dot{f}_s + F_{ab} f_a.$$

If there are no sensor faults and F_{ab} is full column rank, F_{ab} has a left inverse $= (F_a)^L$, and the actuator faults can actually be identified directly as $f_a = (F_a)^L (r)_{eq}$.

6.2 Residual Generation Using SMFO for Subsystem with Unknown Input

Here we examine how fault information in the x_d subsystem can be extracted, even though the x_d subsystem is affected by unknown input. In Section 3 the following SMO was built for x_{id} subsystems

$$\begin{aligned} \dot{\hat{x}}_1^i &= \hat{x}_2^i + L_1^i y_d + B_1^i u + \lambda_1^i sign(y_{id} - \hat{x}_1^i) \\ \dot{\hat{x}}_2^i &= \hat{x}_3^i + L_2^i y_d + B_2^i u + \lambda_2^i sign(r_2^i) \\ &\cdots \quad \cdots \\ \dot{\hat{x}}_{q_i-1}^i &= \hat{x}_{q_i}^i + L_{q_i-1}^i y_d + B_{q_i-1}^i u + \lambda_{q_i-1}^i sign(r_{q_i-1}^i) \\ \dot{\hat{x}}_{q_i}^i &= E_{ia}\hat{x}_a + E_{ib}\hat{x}_b + E_{ic}\hat{x}_c + \textstyle\sum_{j=1}^{m_d} E_{ij}\hat{x}_{jd} + B_{q_i}^i u + \lambda_{q_i}^i sign(r_{q_i}^i) \end{aligned} \tag{70}$$

where $\hat{x}_a, \hat{x}_b, \hat{x}_c$ are estimations for states x_a, x_b, x_c respectively, and

$$r_k^i = (\lambda_{k-1}^i sign(r_{k-1}^i))_{eq} \tag{71}$$

for $k = 2, ..., q_i$, and $r_1^i = y_{1d} - \hat{x}_1^i$ and can be obtained directly. The $\hat{x}_a, \hat{x}_b, \hat{x}_c$ are given by

$$\begin{aligned} \dot{\hat{x}}_a &= A_a \hat{x}_a + L_{ad} y_d + L_{ab} y_b + B_a u, \\ \dot{\hat{x}}_b &= A_b \hat{x}_b + L_{bd} y_d + B_b u + K_b y_b, \\ \dot{\hat{x}}_c &= A_c \hat{x}_c + L_{cd} y_d + L_{cb} y_b + G_c E_{ca} \hat{x}_a + B_c u. \end{aligned} \tag{72}$$

We can make $\hat{x}_b \to x_b$ by designing K_b such that $A_b - K_b C_b$ is a stable matrix. However, $e_a = x_a - \hat{x}_a$ will not be zero if A_a is an unstable matrix and the initial estimation error is nonzero. Also, $e_c = x_c - \hat{x}_c$ will not be zero because of the unknown input. However, if no sensor faults appear and

$$\lambda_k^i > \|x_{k+1}^i - \hat{x}_{k+1}^i\|$$

$k = 1, 2, ..., q_i - 1$, as well as

$$\lambda_{q_i}^i > \|E_{ia} e_a + E_{ic} e_c + d_i\|$$

the observer (70) will give the right estimation of x_{id}.

To analyze the effect of the faults, let us consider the dynamic equation of the estimation error $e_1^i = x_1^i - \hat{x}_1^i$, which is given by

$$\dot{e}_1^i = e_2^i - L_1^i f_{sd} + F_{d1}^i f_a - \lambda_1^i sign(e_1^i + f_{sd}^i)$$

and the residual is given by

$$r_1^i = y_{id} - \hat{x}_1^1 = e_1^i + f_{sd}^i \tag{73}$$

Since $e_1^i = 0$ so long that $f_{sd}^i = 0$, the above algebraic equation for r_1^i implies that r_1^i will become nonzero as f_{sd}^i becomes nonzero. On the other hand, the dynamics of r_1^i is governed by

$$\dot{r}_1^i = e_2^i - L_1^i f_{sd} + \dot{f}_{sd}^i + F_{d1}^i f_a - \lambda_1^i sign(r_1^i). \tag{74}$$

Since detection of incipient faults is of interest, it is expected that magnitude of $-L_1^i f_{sd} + \dot{f}_{sd}^i + F_{d1}^i f_a$ will be small. Note that we already chose $\lambda_1^i > \|x_2^i - \hat{x}_2^i\| = e_2^i$, and it is expected that

$$\|e_2^i - L_1^i f_{sd} + \dot{f}_{sd}^i + F_{d1}^i f_a\| < \lambda_1^i.$$

Under the above condition, r_1^i will approximate zero. With $r_1^i = 0$, its equivalent control signal will be

$$r_2^i = (\lambda_1^i sign(r_1^i))_{eq} = e_2^i - L_1^i f_{sd} + \dot{f}_{sd}^i + F_{d1}^i f_a. \tag{75}$$

r_2^i is used for the estimation of x_2^i, therefore,

$$\dot{e}_2^i = e_3^i - L_2^i f_{sd} + F_{d2}^i f_a - \lambda_2^i sign(r_2^i).$$

and the dynamics of r_2^i will be governed by

$$\dot{r}_2^i = e_3^i - L_2^i f_{sd} - L_1^i \dot{f}_{sd} + \dot{f}_{sd}^i + F_{d2}^i f_a + F_{d1}^i \dot{f}_a - \lambda_2^i sign(r_2^i).$$

As long as λ_2^i is large enough, r_2^i will approximate zero. However, if we compare the algebraic representations of r_1^i and r_2^i, namely equations (73) and (75), we note that the derivative of sensor fault f_{sd}^i appears in r_2^i, which is much more significant than f_{sd}^i itself in r_1^i.

The equivalent control signal of r_2^i is used further for the estimation of x_3^i, and so on, until $r_{q_i}^i$ is calculated. Following the above derivation process and so long that no fault is large enough to make the system sliding stop, we have following the following algebraic description of $r_k^i(k = 1, ..., q_i)$,

$$r_k^i = e_k^i - \sum_{j=1}^{k-1} L_j^i f_{sd}^{(k-j-1)} + \sum_{j=1}^{k-1} F_{dj}^i f_a^{(k-j-1)} + (f_{sd}^i)^{(k-1)} = e_k^i + f_k^i,$$

where $(\cdot)^j$ refers to the the jth derivative. On the other hand, all $r_k^i(1 \leq k \leq q_i, 1 \leq i \leq m_d)$ will dynamically approximate zero. However, we may observer peak phenomenon in $r_k^i(2 \leq k)$ if the derivative of the fault signal changes abruptly from zero to a nonzero value, which is likely to be the case for incipient faults. For large ks a higher order of the derivative appears in r_k^i which will make fault detection possible. At the same time, fault isolation is possible if the fault distribution matrices $L_k^i, F_{dk}^i(k = 1, ..., q_i, i = 1, ..., m_d)$ have some special structure such that r_k^i is only sensitive to certain faults.

The drawback of fault detection based on residual r_k^i is that it is not possible to distinguish between sensor or actuator jitter and more persistent sensor or actuator faults. Again this is so because both events will produce peak phenomena. On the other hand no other effect such as system uncertainties or unknown inputs will produce such a peak. Therefore, occurrence of peaks in r_k^i can provide useful information for fault detection.

Remark 11. We can derive the equivalent control signal of $r_{q_i}^i$ as

$$\begin{aligned} r_{q_i+1}^i &= (\lambda_{q_i}^i sign(r_{q_i}^i))_{eq} \\ &= E_{ia} e_a + E_{ib} e_b + E_{ic} e_c + \textstyle\sum_{j=1}^{m_d} E_{ij} e_{jd} + \dot{f}_{q_i} + F_{dq_i}^i f_a + d_i. \end{aligned}$$

Even under healthy conditions, $r_{q_i+1}^i$ will be nonzero because of the estimation errors of e_a, e_c and the unknown input d_i. We may also observer peak phenomenon in $r_{q_i+1}^i$ because it contains the derivative of faults. However, since it is never known how d_i changes, an occurrence of a peak phenomenon in $r_{q_i+1}^i$ is an unreliable means for fault detection.

Remark 12. Note that if $q_i = 1$ for all $i = 1, ..., m_d$, the above SMO is reduced to SMOO for x_d subsystem. There is no way to distinguish between the faults and the unknown inputs under this condition.

In summary, to accomplish robust fault diagnosis for a linear uncertain system, it is proposed to decompose it into the SCB form. In the SCB form since x_a and x_c subsystems have no outputs, they can not be used for the purpose of residual generation. If x_b subsystem exists, the residual generation discussed in section 6.1 can be applied. If x_d subsystem exist and not all $q_i = 1, (i = 1, ..., m_d)$, we can build SMFO for x_d subsystem and generate the residual (71).

It should be stressed that in practice if the number of independent unknown inputs is larger than the number of independent outputs, no unknown input free subsystem exists and many of the commonly used techniques such as the UIO based FDI will not be applicable. In such cases the SMFO for x_d subsystem significantly improves the possibility of robust fault detection for the linear uncertain systems.

7 SMFO Based Incipient Fault Diagnosis for Nonlinear Uncertain Systems

We consider the nonlinear uncertain systems described as:

$$\begin{aligned} \dot{x} &= A(x) + B(x,u) + F_a(x,u)f_a + \textstyle\sum_{i=1}^{m} g_i(x)d_i(x,u,t) \\ y &= H(x) + f_s \end{aligned} \tag{76}$$

where $H(x) = [h_1(x), ..., h_p(x)]^T$. In Section 4 it was shown that if the nonlinear system (76) satisfies the following conditions:

i. $p \geq m$, where p, m are dimensions of the output and the unknown inputs respectively,
ii. First m outputs have the relative degree $\{q_1, ..., q_m\}$ corresponding to the unknown input matrix $G(x) = [g_1(x), ..., g_m(x)]$, and the distribution spanned by the vector fields $g_1(x), ..., g_m(x)$ is involutive,

then there exists a nonlinear transformation $\Phi_o(x)$ such that (76) can be decomposed into one subsystem x_d with unknown inputs, and one subsystem without unknown inputs. The x_d subsystem can be further decomposed into m subsystems with triangular structure, or $x_d = [x_{1d}, ..., x_{md}]^T$, and each x_{id} subsystem has q_i states, or $x_{id} = [x_1^i, ..., x_{q_i}^i]^T, i = 1, ..., m$. The equations are

$$\begin{aligned} \dot{x}_1^i &= x_2^i + b_1^i(x_d, x_o, u) + F_{d1}^i(x,u)f_a \\ \cdots & \quad \cdots \\ \dot{x}_{q_i-1}^i &= x_{q_i}^i + b_{q_i-1}^i(x_d, x_o, u) + F_{d(q_i-1)}^i(x,u)f_a \\ \dot{x}_{q_i}^i &= a_i(x_d, x_o) + b_{q_i}^i(x_d, x_o, u) + \textstyle\sum_{j=1}^{m} c_{ij}(x_d, x_o)d_j + F_{dq_i}^i(x,u)f_a \\ y_{id} &= x_1^i + f_{sd}^i \end{aligned} \tag{77}$$

The unknown input free subsystem can be written as

$$\begin{aligned} \dot{x}_o &= q(x_d, x_o) + p(x_d, x_o, u) + s(x_d, x_o, u, f_a) \\ y_{m+1} &= h_{m+1}(x_d, x_o) + F_{m+1}(x)f_s \\ \cdots & \quad \cdots \\ y_p &= h_p(x_d, x_o) + F_p(x)f_s \end{aligned} \tag{78}$$

In Section 4 it was also shown that if the input term of subsystem (77) is in the following form

$$\begin{array}{ll} b_1^i(x_d, x_o, u) & = b_1^i(y_r, u) \\ b_2^i(x_d, x_o, u) & = b_2^i(x_2^i, y_r, u) \\ \dots & \dots \\ b_{q_i-1}^i(x_d, x_o, u) & = b_{q_i-1}^i(x_2^i, \dots, x_{q_i-1}^i, y_r, u) \end{array} \tag{79}$$

where y_r represents the sensor fault-free output, a nonlinear SMO can be built because of its triangular structure,

$$\begin{array}{ll} \dot{\hat{x}}_1^i & = \hat{x}_2^i + b_1^i(y, u) + \lambda_1^i sign(y_{id} - \hat{x}_1^i) \\ \dot{\hat{x}}_2^i & = \hat{x}_3^i + b_2^i(\hat{x}_2^i, y, u) + \lambda_2^i sign(r_2^i) \\ \dots & \dots \\ \dot{\hat{x}}_{q_i-1}^i & = \hat{x}_{q_i}^i + b_{q_i-1}^i(\hat{x}_2^i, \dots, \hat{x}_{q_i-1}^i, y, u) + \lambda_{q_i-1}^i sign(r_{q_i-1}^i) \\ \dot{\hat{x}}_{q_i}^i & = a_i(\hat{x}_d, \hat{x}_o) + b_{q_i}^i(\hat{x}_d, \hat{x}_o, u) + \lambda_{q_i}^i sign(r_{q_i}^i) \end{array} \tag{80}$$

where $r_k^i (k = 2, \dots, q_i)$ are given by (71). In this case r_k^i can be written as

$$r_k^i = e_k^i + \sum_{j=1}^{k-1} (\Delta b_j^i)^{(k-j-1)} + \sum_{j=1}^{k-1} (F_{dj}^i(x, u) f_a)^{(k-j-1)} + (f_{sd}^i)^{(k-1)} = e_k^i + f_k^i$$

where

$$\Delta b_j^i = b_j^i(x_2^i, \dots, x_j^i, y_r, u) - b_j^i(\hat{x}_2^i, \dots, \hat{x}_j^i, y_r + f_s, u).$$

Finally, the $r_{q_i+1}^i = (\lambda_{q_i}^i sign(r_{q_i}^i))_{eq}$ is equal to

$$\begin{array}{rl} r_{q_i+1}^i = & a_i(x_d, x_o) - a_i(\hat{x}_d, \hat{x}_o) + b_{q_i}^i(x_d, x_o, u) - b_{q_i}^i(\hat{x}_d, \hat{x}_o, u) \\ & + \sum_{j=1}^m c_{ij}(x_d, x_o) d_j + F_{dq_i}^i(x, u) f_a + \dot{f}_{q_i}. \end{array} \tag{81}$$

Similar to the linear case, the usage of $r_k^i (2 \le k \le q_i)$ for fault detection is based on a peak phenomenon which would arise due to derivative of fault signal.

To exploit the usage of an unknown input free subsystem for FDI, one can try to build an observer for the x_o subsystem. If x_o is an observable system and an observer could be built for it, then the following residuals can be formulated:

$$r_j = h_j(\hat{x}_d, \hat{x}_o) - y_j, j = m+1, \dots, p.$$

Unfortunately, the complete solution for nonlinear systems in the form of (78) is an open problem and needs to be studied in the future.

Here we propose a sliding mode output functional observer (SMOFO) and discuss its use for incipient fault diagnosis. Assume

$$y_{r,j} = h_j(x_d, x_o), \qquad j = m+1, \dots, p$$

$$Y(x) = [y_{r,m+1}, ..., y_{r,p}]^T$$

$$f_{ts} = [F_{m+1}(x)f_s, ..., F_p(x)f_s]$$

and

$$Y_o = [y_{m+1}, ..., y_p]^T$$

Obviously $Y_o = Y(x) + f_{ts}$ and $Y_o = Y(x)$ when $f_s = 0$. The following SMOFO is proposed

$$\dot{z} = L(Y_o, \hat{x}_d, u) + \Lambda sign(W(Y_o) - z). \tag{82}$$

It is expected that $z \to W(Y(x))$. Let $e = W(Y(x)) - z$, in which case

$$\begin{aligned}\dot{e} &= \tfrac{\partial W(Y_o)}{\partial Y_o}\tfrac{\partial Y_o}{\partial x_o}(q(x_d, x_o) + p(x_d, x_o, u) + s(x_d, x_o, u, f_a)) \\ &\quad - L(Y_o, \hat{x}_d, u) - \Lambda sign(W(Y_o) - z) \\ &= M(x_d, x_o, u, f_a) - L(Y_o, \hat{x}_d, u) - \Lambda sign(W(Y_o) - z).\end{aligned}$$

Set $r = W(Y_o) - z = W(Y(x) + f_{ts}) - z$, then $W(Y(x) + f_{ts})$ can be represented in a Taylor series expansion as

$$W(Y(x) + f_{ts}) = W(Y(x)) + \sum_{k=1}^{\infty} \frac{W(Y(x))^{(k)} f_{ts}^k}{k!} = W(Y(x)) + Q(x, f_{ts})$$

where $W(Y(x))^{(k)}$ refers to the kth derivative. Thus, $r = e + Q(x, f_{ts})$. Representing the derivative of $Q(x, f_{ts})$ as

$$\frac{\partial Q}{\partial x}\dot{x} + \frac{\partial Q}{\partial f_{ts}}\dot{f}_{ts} = D(x, f_{ts}, \dot{f}_{ts}).$$

then

$$\dot{r} = M(x_d, x_o, u, f_a) - L(Y_o, \hat{x}_d, u) - D(x, f_{ts}, \dot{f}_{ts}) - \Lambda sign(r).$$

Similar to the principle of SMOO, r will approximate zero due to the high gain design of Λ and will stay at zero even when there are incipient faults. However, we can set residual for fault detection as the equivalent control signal of r, which is

$$r_{eq} = (\Lambda sign(r))_{eq} = M(x_d, x_o, u, f_a) - L(Y_o, \hat{x}_d, u) - D(x, f_{ts}, \dot{f}_{ts}). \tag{83}$$

Now to accomplish fault detection based upon the above, the following need to be satisfied:

i. $r_{eq} = 0$ when there are no faults, namely, $f_s = f_a = 0$. It is easy to show that $Q(x, f_{ts}) = 0$ and $D(x, f_{ts}, \dot{f}_{ts}) = 0$ if there are no faults. Therefore, r_{eq} is reduced to

$$r_{eq} = M(x_d, x_o, u, 0) - L(Y(x), \hat{x}_d, u).$$

Note that $\hat{x}_d \rightarrow x_d$ when there are no faults, thus x_d can be considered as a known input for the x_o subsystem. Given this, the objective of having $r_{eq} = 0$ under fault free condition can be achieved if and only if

$$M(x_d, x_o, u, 0) = L(Y(x), x_d, u). \tag{84}$$

ii. r_{eq} is nonzero when $f_a \neq 0$ or $f_s \neq 0$. In practice this property need to be checked using the equation (83).

Remark 13. Compared with SMOO, the SMOFO introduces one more degree of freedom in the design process. That is the function of output $W(Y(x))$, which could allow the effect of unknown states variable x_o can be removed through nonlinear state transformation. This design avoids the complex problem of estimating states for nonlinear systems. Unfortunately, we still need to find a general means for designing $W(Y(x))$ and $L(Y(x), x_d, u)$ such that condition (84) is satisfied. In the next section it will be shown that solutions could be found through trial and error for certain nonlinear systems with special structure.

8 Illustrative Examples

Example 1. In this example we shall briefly outline the application of SMOs in engine diagnostics application. This example is concerned with the diagnostics issues on one engine subsystem, namely, the intake manifold.

In this study a model of the intake manifold dynamics was obtained based on the physics of the intake process. This model can be used to generate engine variables such as the manifold pressure (MAP), and mass air flow (MAF) based on the throttle input, Exhaust Gas Recirculation(EGR), and RPM inputs. It is possible then to use the error between the measured MAP and the predicted MAP from the model to detect MAP sensor failure. It can be shown that for the particular case of the manifold dynamics which is a very stable dynamic system, this approach can work for estimating the intake manifold variables. However, in general it is well known that the dynamics of the open loop model can not be used for estimation purposes as the estimates may be unstable or slow to converge to their true values. In addition, since model variation due to engine aging and car to car variations are expected, it is important to consider improvements to the open loop strategy. Therefore, it was also the purpose of this study to investigate means for robustifying the model based diagnosis methodology. Here we investigate the design of a robust estimation and diagnosis scheme using a sliding mode observer strategy.

The model used for this study is based on a V8 engine model that was developed and validated (excluding the EGR model) using data from an experimental vehicle [44]. The model used for this example was build in XMATH's System Build environment and is briefly discussed here.

Throttle Body

$$m_a = \frac{(C_d A_{th} + C_{di} A_i) P_0}{\sqrt{RT_0}} \Psi \tag{85}$$

where

$$\Psi = \begin{cases} \sqrt{\frac{2k}{k-1}((\frac{P}{P_0})^{\frac{2}{k}} - (\frac{P}{P_0})^{\frac{k+1}{k}})} & \text{if} \quad \frac{P}{P_0} > (\frac{2}{k+1})^{\frac{k}{k-1}} \\ \sqrt{k(\frac{2}{k+1})^{\frac{k+1}{k-1}}} & \text{otherwise} \end{cases}$$

m_a: Air flow rate
P: manifold pressure
P_0: atmospheric pressure
k: ratio of specific heats
A_{th}: cross sectional area of throttle
A_i: idle Air Control (IAC) valve area
C_d: orifice discharge coefficient
R: Ideal gas constant for air
T_0: Ambient temperature

The throttle flow area is a function of throttle geometry and throttle angle,

$$A_{th} = c_1 + c_2(1 - \cos(\theta - \theta_0))$$

Intake Manifold Dynamic

$$\dot{P} = \frac{R_m T_m}{V}(m_a + m_{egr} - m_e) \tag{86}$$

$$\dot{r} = \frac{(1+r) R_m T_m}{PV}(m_{egr} - r m_t) \tag{87}$$

where

R_m: Ideal gas constant
T_m: Temperature
r: Fraction of EGR
m_e: Mass flow rate into engine cylinders
m_{egr}: EGR flow rate
V: Intake manifold volume

$$m_e \triangleq \frac{\eta_v V_d P \omega}{4\pi R_m T_m} \tag{88}$$

where

$$R_m = \frac{1}{r+1} R + \frac{r}{1+r} R_e \quad T_m = \frac{1}{1+r} T_0 + \frac{r}{1+r} T_e$$

V_d : engine displacement
ω : engine speed
T_0 : ambient temperature
T_e : EGR temperature
R_e: Ideal gas constant for EGR
η_v : Volumetric efficiency (function of MAP & ω via look-up table).

Rotational Dynamics and Torque Converter

$$\dot{\omega} = \frac{1}{J}(T_i - T_f - T_p - T_L) \tag{89}$$

$$T_p = C_1\omega^2 - C_2\omega\omega_t - C_3\omega_t^2 \tag{90}$$

where
T_i: Indicated torque
T_f: Friction torque
T_L: Load torque
T_p: Torque converter pump torque
J: Engine inertia
ω_t: Torque converter's turbine speed
$C_i, i = 1, 2, 3$: Constants

As mentioned before, in this study EGR effect was not modeled, i.e., $r = 0$, so that from (85)-(88), we have the following,

$$\begin{aligned}\dot{P} &= \frac{RT_0}{V}(m_a - m_e)\\ &= \frac{RT_0}{V}\left[\frac{C_d(c_1 + c_2(1 - \cos(\theta - \theta_0))) + C_{di}A_i}{\sqrt{RT_0}}P_0\Psi(P) - \frac{\eta_v V_d P\omega}{4\pi RT_0}\right]\end{aligned} \tag{91}$$

For notational convenience set

$$x \triangleq P$$

combining (89)-(91), we can rewrite the dynamic model for x,

$$\dot{x} = -a(\omega)x + b(\theta, t)\Psi(x) \tag{92}$$

where the output equation is given by $y = x$. Where

$$a(\omega) \triangleq \frac{\eta_v V_d \omega}{4\pi V} \qquad b(\theta, t) \triangleq \frac{C_d(c_1 + c_2(1 - \cos(\theta - \theta_0))) + C_{di}A_i}{V}\sqrt{RT_0}P_0$$

Both $a(\omega)$ and $b(\theta, t)$ are positive functions, and from our model, we estimate

$$a(\omega) \leq 0.0531\omega, \qquad b(\theta, t) \leq 9000000.$$

Assuming that (92) represent real dynamics of the throttle body plus the intake manifold subsystems, with the knowledge of the input variables (e.g., speed, throttle angle, etc) it is possible to use a similar model along an assumed initial condition on the manifold pressure to estimate future values of the manifold pressure. This is possible, by using the following model,

$$\dot{\hat{x}} = -a(\omega)\hat{x} + b(\theta, t)\Psi(\hat{x}) \tag{93}$$

Note that any possible difference between the outputs of (92) and (93) is due to the initial condition. Set

$$\tilde{x} = x - \hat{x}$$

then the error dynamics becomes,

$$\dot{\tilde{x}} = -a(\omega)\tilde{x} + b(\theta, t)(\Psi(x) - \Psi(\hat{x})) \tag{94}$$

we know that

$$\Psi(x) - \Psi(\hat{x}) = \Psi'(\bar{x})\tilde{x}, \tag{95}$$

where $\bar{x}$ is between x and $\hat{x}$.

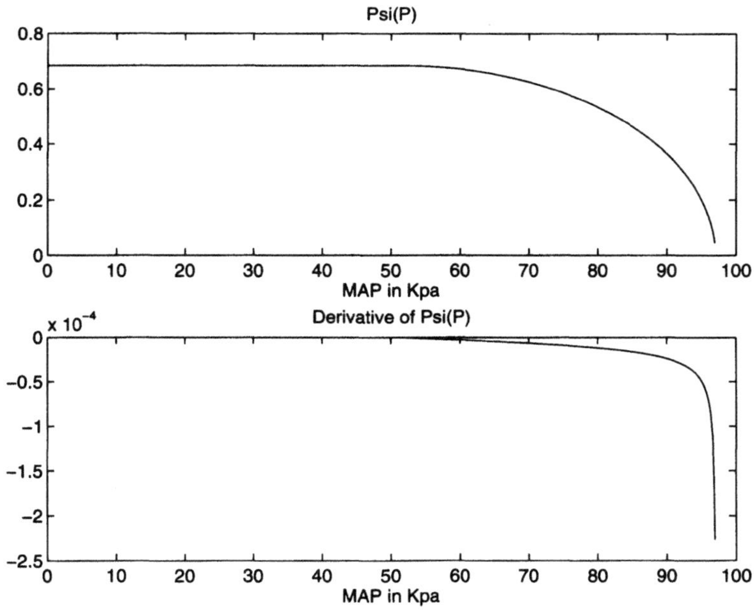

Fig. 3. $\Psi(P)$ and $\Psi'(P)$

Combine (94) and (95), we have,

$$\dot{\tilde{x}} = (-a(\omega) + b(\theta, t)\Psi'(\bar{x}))\tilde{x}$$

and from Figure 3, $\Psi' \leq 0$, so $-a + b\Psi' < 0$, then the above system is asymptotically stable. This is the main reason that one can theoretically use the open loop model of this particular system to estimate its state, even with a wrong initial condition. On the other hand, when there are some deviation between the plant and the model, then this open-loop model-based estimation will not yield accurate estimates and experimental data from two cars with the same engine class indicated this as well.

As a result, for the intake manifold dynamics under investigation, we propose an observer of the form

$$\dot{\hat{x}} = -a(\omega)\hat{x} + b(\theta,t)\Psi(\hat{x}) + K(x-\hat{x}) + K_0\mathrm{sgn}(x-\hat{x}) \tag{96}$$

Given the above observer, the error dynamics of the estimation and plant state is,

$$\dot{\tilde{x}} = (-a(\omega) - K)\tilde{x} + b(\theta,t)(\Psi(x) - \Psi(\hat{x})) + d - K_0\mathrm{sgn}(x-\hat{x}) \tag{97}$$

By using an argument based upon the Lyapunov approach, we can show that as long as the disturbances and uncertainties are bounded, i.e., $d < K_0$, then the sliding mode observer should perform satisfactorily.

Figure 4 shows actual data from a test vehicle during a typical city driving cycle. Figure 5 illustrates the comparison between the open loop and the sliding mode observer's estimate of the MAP when there are plant uncertainties present in the system. In Figure 6, the estimation error is illustrated which clearly indicates that the sliding mode observer can provide a more robust estimate. Finally, Figure 7 illustrates the fact that the MAP sensor failures can still be detected when using the sliding mode observer whereas in the case of the open loop observer, if there are model mismatches present due to car to car variations, then this would be impossible.

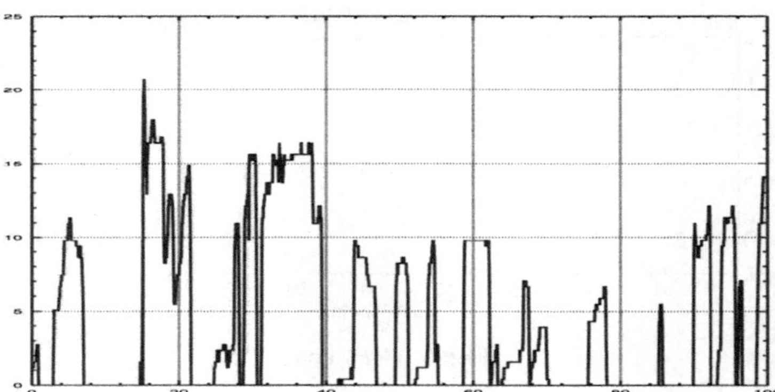

Fig. 4. Throttle input

Example 2. The system under consideration is a one-link manipulator with revolute joints actuated by a DC motor as shown in Figure 8. The elasticity

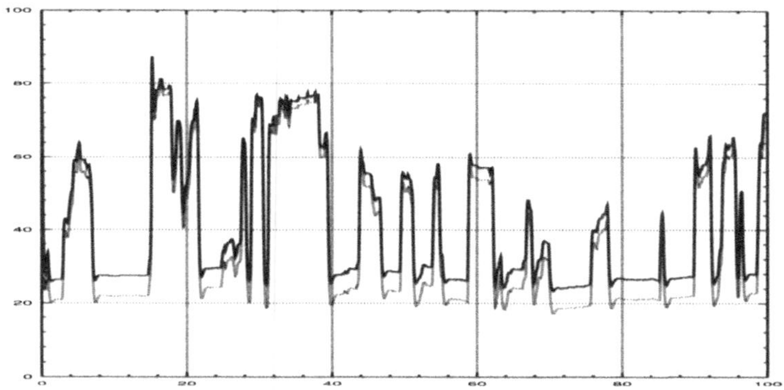

Fig. 5. Actual, open loop, and closed loop MAP estimates with plant uncertainty.The closed loop estimator practically provides an estimate that is very close to the actual MAP reading. On the other hand, the open loop estimator clearly can not provide a good estimate due to the model mismatch.

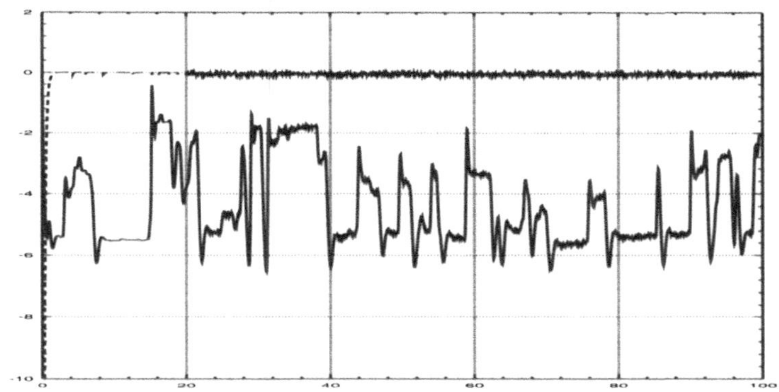

Fig. 6. Residuals of the open and closed loop observers with $d(t) = 40,000$.The closed loop estimator's superior performance can be seen from the residuals.

of the joint can be well-modelled by a linear torsional spring [50]. The elastic coupling of the motor shaft to the link introduces an additional degree of freedom. The states of this system are motor position and velocity, and the link position and velocity.

The corresponding state-space model is

$$\begin{aligned} \dot{\theta_m} &= \omega_m \\ \dot{\omega_m} &= \tfrac{k}{J_m}(\theta_l - \theta_m) - \tfrac{B}{J_m}\omega_m + \tfrac{K_r}{J_m}u \\ \dot{\theta}_l &= \omega_l \\ \dot{\omega}_l &= -\tfrac{k}{J_l}(\theta - \theta_m) - \tfrac{mgh}{J_l} sin(\theta_l) \end{aligned}$$

with J_m being the inertia of the motor; J_l being the inertia of the link; θ_m the angular rotation of the motor; θ_l the angular position of the link; ω_m the

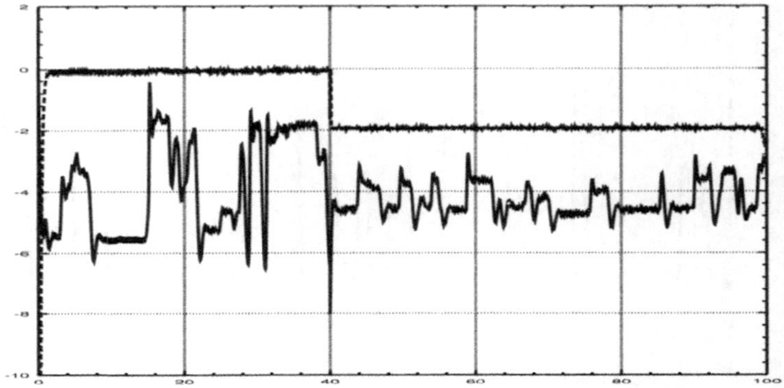

Fig. 7. MAP residuals for open and close loop observer with MAP sensor failure at $t = 20$ seconds. ($d(t)$ changing)

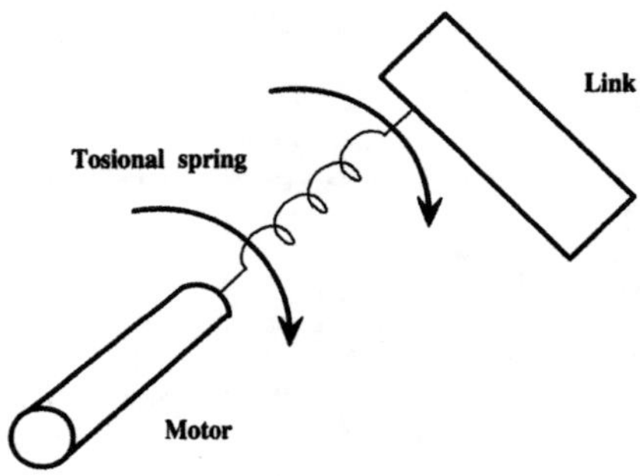

Fig. 8. A Flexible Link Robot

angular velocity of the motor; and ω_l the angular velocity of the link. Thus, the system dynamics are nonlinear and of the form

$$\dot{x} = Ax + \Phi(x) + Bu$$

with $x = [\theta_m, \omega_m, \theta_l, \omega_l]^T$

$$A = \begin{bmatrix} 0 & 1 & 0 & 0 \\ -48.6 & -1.25 & 48.6 & 0 \\ 0 & 0 & 0 & 1 \\ 19.5 & 0 & -19.5 & 0 \end{bmatrix}, B = \begin{bmatrix} 0 \\ 21.6 \\ 0 \\ 0 \end{bmatrix}, \Phi(x) = \begin{bmatrix} 0 \\ 0 \\ 0 \\ -3.33 sin(x_3) \end{bmatrix}$$

The above parameters for the system are typical and have been taken from [50]. A nonlinear input–output linearizing control law for this system is presented in [50]. The control law guarantees closed-loop stability and tracking of any desired trajectory by the robot link. However, this control law requires measurement of all the states, and the measurement of angular position (x_3) and velocity (x_4) of the link is difficult. Reference [40] proposed an observer by considering this system as a Lipschitz nonlinear system, where motor position and velocity are measured. Obviously, this systems can be considered as a linear system subject to bounded unknown input, namely the nonlinear term $-3.33sin(x_3)$. Further it is easy to verify that even if only motor position is measured, or

$$y = \begin{bmatrix} 1\,0\,0\,0 \end{bmatrix} x,$$

a sliding mode observer can be designed using the proposed SMO algorithm. The observer based design will relieve the need for measurement of motor velocity. It should be stressed that no Walcott-Zak SMO or linear UIO for the above robotic system exists because $rank(CG) = 0 < rank(G) = 1$. The transformation

$$\Gamma_1 = \begin{bmatrix} 1 & 0 & 0 & 0 \\ 0 & 1 & 0 & 0 \\ 1 & 0.0257 & 0.0206 & 0 \\ 0 & 1 & 0.0257 & 0.0206 \end{bmatrix}; \Gamma_2 = 1; \Gamma_3 = 0.0206;$$

will transform the system into

$$\dot{x} = \begin{bmatrix} 0 & 1 & 0 & 0 \\ 0 & 0 & 1 & 0 \\ 0 & 0 & 0 & 1 \\ 0 & -24.375 & -68.1 & -1.25 \end{bmatrix} x + \begin{bmatrix} 0 \\ 21.6 \\ -27.0 \\ 1016.0 \end{bmatrix} u + \begin{bmatrix} 0 \\ 0 \\ 0 \\ 1 \end{bmatrix} d$$

and C is the same as before. The observer is

$$\dot{\hat{x}} = \begin{bmatrix} 0 & 1 & 0 & 0 \\ 0 & 0 & 1 & 0 \\ 0 & 0 & 0 & 1 \\ 0 & -24.375 & -68.1 & -1.25 \end{bmatrix} \hat{x} + \begin{bmatrix} 0 \\ 21.6 \\ -27.0 \\ 1016.0 \end{bmatrix} u + \begin{bmatrix} \lambda_1 sign(y_1 - \hat{x_1}) \\ \lambda_2 sign(\overline{e_2}) \\ \lambda_3 sign(\overline{e_3}) \\ \lambda_4 sign(\overline{e_4}) \end{bmatrix}$$

Actually, the observer can be designed without a need for SCB transformation. First, build the observer for x_1 as

$$\dot{\hat{x_1}} = \hat{x_2} + \lambda_1 sign(y_1 - \hat{x_1})$$

After e_1 has approximately reached zero, we know $e_2 = (\lambda_1 sign(e_1))_{eq}$. Next, build the observer for x_2 as

$$\dot{\hat{x_2}} = -48.6y_1 - 1.25\hat{x_2} + 48.6\hat{x_3} + \lambda_2 sign(e_2)$$

After e_2 approximates to zero, we know that

$$e_3 = (\lambda_1 sign(e_2))_{eq}/48.6$$

and the observer for x_3 is built as

$$\dot{\hat{x_3}} = \hat{x_4} + \lambda_3 sign(e_3)$$

Finally, $e_4 = (\lambda_3 sign(e_3))_{eq}$ and the observer for x_4 is

$$\dot{\hat{x_4}} = -19.5y_1 + 19.5\hat{x_4} + \lambda_4 sign(e_4)$$

e_4 will go zero in spite of the nonlinear term $-3.33sin(x_3)$. Actually, it does not matter if the nonlinear term is more complicated, Lipschitz or not. In simulation, we use the saturation function to replace sign function and upper limit is set to be 0.02. The switching gains are $\lambda_1 = 800, \lambda_2 = 120, \lambda_3 = 120$ and $\lambda_4 = 1400$. The equivalent control signal is applied to the second, third and fourth state estimation at $t_1 = 0.2sec, t_2 = 0.8sec$ and $t_3 = 1.8sec$ respectively, the initial states are assumed to be

$$x_0 = \begin{bmatrix} 1 \; 1 \; 1 \; 1 \end{bmatrix}, \hat{x}_0 = \begin{bmatrix} 0 \; 0 \; 0 \; 0 \end{bmatrix}$$

Figure 9 shows that all estimated states converge rapidly to the correct values. We conclude that the proposed SMO is a practical solution to an important robotic application.

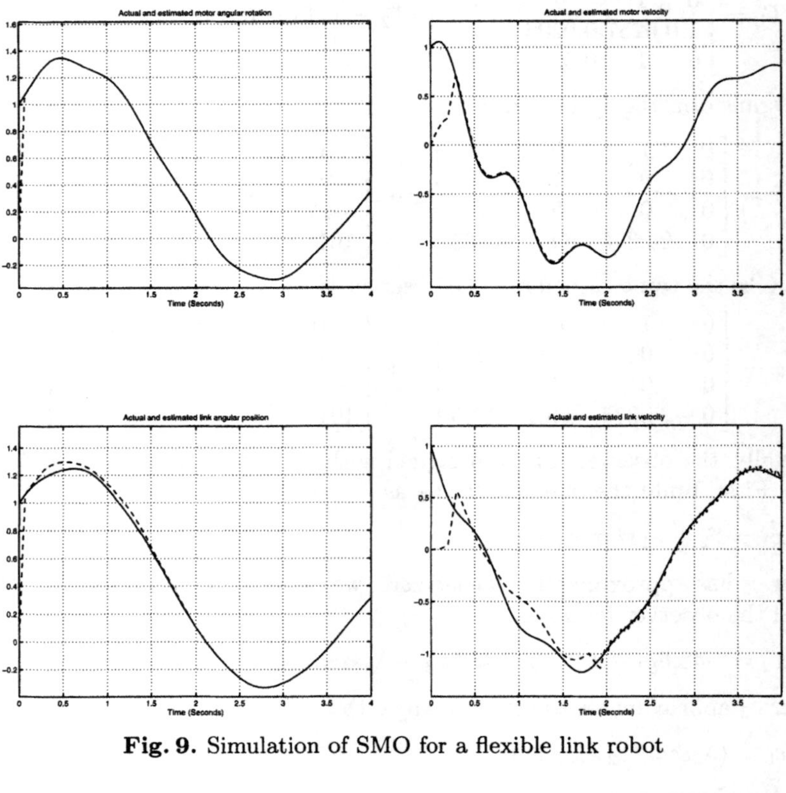

Fig. 9. Simulation of SMO for a flexible link robot

Example 3. The mechatronic system considered is an inverted pendulum. This system has four states: x_1, the position of base; $x_3 = \dot{x}_1$, the velocity of the base; x_2, the angular position of the pendulum; $x_4 = \dot{x}_2$, the angular velocity of the pendulum. It is assumed that x_1, x_2 and x_3 are measurable.

The input variable is the input voltage, u, to the power amplifier which drives the motor. This is a nonlinear unstable system which is stabilized by an observer-based feedback controller [17]. The linearized closed-loop system can be described as follows:

$$\begin{aligned} \dot{x} &= (A + \Delta A)x + Bu + Br_{ref} + B\xi \\ y &= Cx + f_s \end{aligned} \tag{98}$$

where $u = -K\hat{x}$, and $\hat{x}$ is the estimation of the state x. It also incorporates the system input r_{ref}, the desired position of the moving base, into the amplifier input. ξ represents the effects of the nonlinear friction in the drive train on the pendulum motion. The numerical values of the system matrix A, B, C and the controller gain K are,

$$A = \begin{bmatrix} 0 & 0 & -1.399 & 0 \\ 0 & 0 & 0 & 1 \\ 0 & -0.1389 & -0.546 & 0.001905 \\ 0 & 21.7 & 6.236 & -0.2902 \end{bmatrix}, B = \begin{bmatrix} 0 \\ 0 \\ -4.192 \\ 47.82 \end{bmatrix}, K = \begin{bmatrix} -3.7219 \\ 3.615 \\ 4.7994 \\ 0.7849 \end{bmatrix}^T,$$

$$C = \begin{bmatrix} 1\,0\,0\,0 \\ 0\,1\,0\,0 \\ 0\,0\,1\,0 \end{bmatrix}.$$

The matrix ΔA is unknown, which represents model uncertainty due to the system's nonlinearity. This can be expressed by

$$\Delta A = \begin{bmatrix} 0 & 0 & 0 & 0 \\ 0 & 0 & 0 & 0 \\ 0 & \Delta a_{32} & \Delta a_{33} & \Delta a_{34} \\ 0 & \Delta a_{42} & \Delta a_{43} & \Delta a_{44} \end{bmatrix}.$$

All system uncertainty and disturbance can be grouped together as Gd, where

$$Gd = \begin{bmatrix} 0\,0 \\ 0\,0 \\ 1\,0 \\ 0\,1 \end{bmatrix} \begin{bmatrix} \Delta a_{32}x_2 + \Delta a_{33}x_3 + \Delta a_{34}x_4 - 4.192\xi_1 \\ \Delta a_{42}x_2 + \Delta a_{43}x_3 + \Delta a_{44}x_4 - 47.82\xi_2 \end{bmatrix}.$$

The states and disturbance ξ are bounded, thus the unknown input d is bounded. For details of controller design we refer the reader to [17]. Because actuator fault vector $F_a = B$ belongs to $Im(G)$, it is not separable from the unknown inputs. The detection and isolation of sensor faults is considered here.

We first note that state x_1 and output y_1 form an unknown input free and observable subsystem,

$$\dot{x}_1 = -1.399x_3; y_1 = x_1,$$

where x_3 is the state for a subsystem subject to the unknown input,

$$\dot{x}_3 = -0.1389x_2 - 0.546x_3 + 0.001905x_4 - 4.192u + d_1; y_3 = x_3.$$

The x_1 is equivalent to the x_b subsystem in SCB form, and all states are measurable. Thus, the Utkin SMO is simplified as SMOO, which is constructed as

$$\dot{z} = -1.399y_3 + \Lambda sign(y_1 - z).$$

If $r = y_1 - z$, we know that

$$\dot{r} = 1.399f_s^3 + \dot{f}_s^1 - \lambda sign(r),$$

and

$$r_{eq} = 1.399f_s^3 + \dot{f}_s^1.$$

Therefore, the residual will be sensitive to faults of sensor 1 and 3. Because the dimension of the residual vector is 1, it is impossible to isolate these two faults.

We note that x_2, x_4 is in the form of a two dimensional x_d subsystem. The following SMFO can be constructed:

$$\begin{aligned}
\dot{\hat{x}}_2 &= \hat{x}_4 + \lambda_1 sign(y_2 - \hat{x}_2) \\
\dot{\hat{x}}_4 &= 21.7\hat{x}_2 - 0.2902\hat{x}_4 + 6.236y_3 + 47.82u + \lambda_2 sign((\lambda_1 sign(y_2 - \hat{x}_2))_{eq})
\end{aligned}$$

Thus, we can detect f_s^2 by the following fact:

$$r_2 = (\lambda_1 sign(y_2 - \hat{x}_2))_{eq} = e_4 + \dot{f}_s^2$$

Figure 10 shows the simulation result of SMFO where sensor noise with maximum magnitude 0.01 is introduced, and $\lambda_1 = 2000, \lambda_2 = 500$. We use the saturation function to replace the sign function, and the upper limit is set to be 0.02. The subplot a) is the shape of incipient sensor fault signal f_s^2, and subplot b) is the residual r_2. It is noted that a peak appears in the residual when the fault occurs.

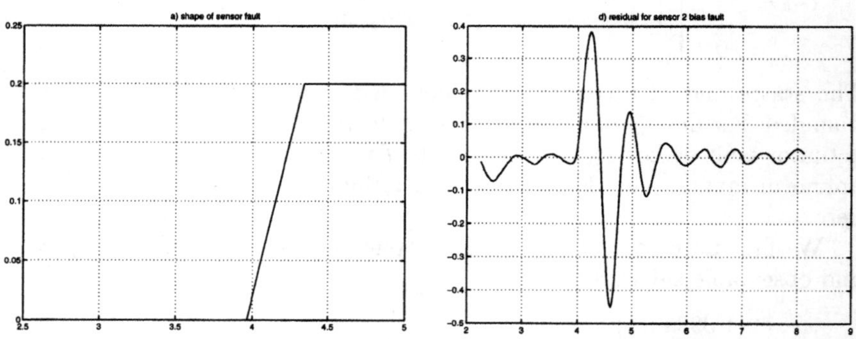

Fig. 10. Fault diagnosis simulation using SMFO for an inverted pendulum

Example 4. A three-phase current motor model [3] is described by the following nonlinear equations:

$$\dot{x} = \begin{bmatrix} f_1(x) \\ f_2(x) \\ f_3(x) \end{bmatrix} + B(x)u + g(x)d$$

$$= \begin{bmatrix} x_2 \\ -A_1x_2 - A_2x_3 sinx_1 - A_3 sin2x_1 \\ -D_1x_3 + D_2 cosx_1 \end{bmatrix} x + \begin{bmatrix} 0\,0 \\ 1\,0 \\ 0\,1 \end{bmatrix} \begin{bmatrix} u_1 \\ u_2 \end{bmatrix} + \begin{bmatrix} 0\,0 \\ 1\,0 \\ 0\,1 \end{bmatrix} \begin{bmatrix} d_1 \\ d_2 \end{bmatrix}$$

where $x = [x_1, x_2, x_3]^T$, x_1, x_2 and x_3 denote the model states rotor angle, speed deviation and field flux linkage, respectively. The known inputs are u_1 (nominal mechanical power input) and u_2 (field voltage), the and unknown inputs are

$$d_1 = \Delta A_1 x_2 + \Delta A_2 x_3 sinx_1 + \Delta A_3 sin2x_1$$

which represent uncertainties of parameters A_1, A_2 and A_3, and

$$d_2 = \Delta D_1 x_3 + \Delta D_2 cosx_1$$

which represents uncertainties of parameters D_1, D_2. All changes are induced by the operating temperature, or component incipient faults. Signals d_1 or d_2 may be small enough to be neglected in different operational conditions. To illustrate the robust SMFO design, we consider several cases.

Case 1: Both d_1 and d_2 are nonzero

In this case, it is noted that x_1, x_2 is a sub-system in the triangular form of (47) if x_1 is measured, namely if $y_1 = x_1$, it will make x_2 observable. If we have $y_2 = x_3$, the all states will be estimated by following observer,

$$\begin{aligned} \dot{\hat{x}}_1 &= \hat{x}_2 + \lambda_1 sign(y_1 - \hat{x}_1) \\ \dot{\hat{x}}_2 &= -A_1\hat{x}_2 - A_2 y_2 sin y_1 - A_3 sin2y_1 + \lambda_2 sign((\lambda_1 sign(y_1 - \hat{x}_1)_{eq}) \\ \dot{\hat{x}}_3 &= -D_1\hat{x}_3 + D_2 cos y_1 + \lambda_1 sign(y_2 - \hat{x}_3) \end{aligned}$$

However, if the second output $y_2 = x_3$ is not available, only a SMFO, but not a SMO, can be designed, because x_3 cannot be estimated correctly under this condition.

Case 2: d_1 is nonzero, d_2 is zero

In this case, x_3 is an unknown input free sub-system. Fortunately, it is detectable because $D_1 > 0$. Therefore, we can build following observer using only one measurement $y_1 = x_1$,

$$\begin{aligned} \dot{\hat{x}}_1 &= \hat{x}_2 + \lambda_1 sign(y_1 - \hat{x}_1) \\ \dot{\hat{x}}_2 &= -A_1\hat{x}_2 - A_2\hat{x}_3 sin y_1 - A_3 sin2y_1 + \lambda_2 sign((\lambda_1 sign(y_1 - \hat{x}_1)_{eq}) \\ \dot{\hat{x}}_3 &= -D_1\hat{x}_3 + D_2 cos y_1 \end{aligned}$$

Case 3: d_1 is zero, d_2 is nonzero

In this case, with only one measurement for x_1, the system can be transformed into triangular form and all states can be estimated. Note that $g(x) = [0 \quad 0 \quad 1]^T$, and the relative degree of output $y = h(x) = x_1$ corresponding to $g(x)$ can be calculated as

$$\frac{\partial h_1}{\partial x} = (0 \quad 0 \quad 1), L_g h_1(x) = 0, L_f h_1(x) = f_1(x) = x_2$$

$$\frac{\partial(L_f h_1)}{\partial x} = \frac{\partial f_1}{\partial x} = (0 \quad 1 \quad 0), L_g L_f h_1(x) = 0; L_f^2 h_1(x) = f_2(x)$$

$$\frac{\partial(L_f^2 h_1)}{\partial x} = (-A_2 x_3 cos x_1 - A_3 cos 2x_1 \quad -A_1 \quad -A_2 sin x_1),$$

$$L_{g_1} L_f^2 h_1(x) = -A_2 sin x_1.$$

Note that $L_g L_f^2 h_1(x) \neq 0$ if $x_1 \neq k\pi$, thus the relative degree is $r_1 = 3$ at point $x_1 \neq k\pi$. This means that we shall be able to find a transformation only locally, away from any point such that $x_1 = k\pi$. The transformation is

$$\begin{aligned}
\xi_1 &= \phi_1(x) = h_1(x) = x_1 \\
\xi_2 &= \phi_2(x) = L_f h_1(x) = x_2 \\
\xi_3 &= \phi_3(x) = L_f^2 h_1(x) = f_2(x).
\end{aligned}$$

The Jacobian matrix of the transformation thus defined

$$\frac{\partial \Phi}{\partial x} = \begin{bmatrix} 1 & 0 & 0 \\ 0 & 1 & 0 \\ -A_2 x_3 cos x_1 - A_3 cos 2x_1 & -A_1 & -A_2 sin x_1 \end{bmatrix}$$

which is nonsingular for all $x_1 \neq k\pi$, and the inverse transformation is given by

$$\begin{aligned}
x_1 &= \xi_1 \\
x_2 &= \xi_2 \\
x_3 &= z(\xi) = \frac{\xi_3 + A_1 \xi_2 + A_3 sin 2\xi_1}{-A_2 sin \xi_1}.
\end{aligned}$$

In these new coordinates the system is described by

$$\begin{aligned}
\dot{\xi}_1 &= \xi_2 \\
\dot{\xi}_2 &= \xi_3 \\
\dot{\xi}_3 &= -A_1 \xi_3 - \xi_2 (A_2 z(\xi) cos \xi_1 + 2A_3 cos 2\xi_1) - A_2 sin \xi_1 (-D_1 z(\xi) + D_2 cos \xi_1) \\
y_1 &= \xi_1.
\end{aligned}$$

Actually, in this example the SMO can be designed without the above complicated transformation calculation. The observer is

$$\begin{aligned}
\dot{\hat{x}}_1 &= \hat{x}_2 + \lambda_1 sign(y_1 - \hat{x}_1) \\
\dot{\hat{x}}_2 &= -A_1 \hat{x}_2 - A_2 \hat{x}_3 sin y_1 - A_3 sin 2y_1 + \lambda_2 sign((\lambda_1 sign(y_1 - \hat{x}_1)_{eq}) \\
\dot{\hat{x}}_3 &= -D_1 \hat{x}_3 + D_2 cos y_1 + \lambda_3 sign(\overline{e_3}).
\end{aligned}$$

Note that the equivalent control signal based on the second equation, is

$$(\lambda_2 sign(\overline{e_2}))_{eq} = -A_2 e_3 siny_1$$

thus

$$\overline{e_3} = \frac{(\lambda_2 sign(\overline{e_2}))_{eq}}{-A_2 siny_1} = e_3.$$

Obviously, it is true only if $siny_1 \neq 0$. The parameters in the model have the value, $A_1 = 0.2703, A_2 = 12.01, A_3 = -48.04, D_1 = 0.3222, D_2 = 1.9$, and $\Delta D_1 = 0.1, \Delta D_2 = 0.6$. The control input $u_1 = 36.19, u_2 = 1.9333$. The gain $\lambda_1 = \lambda_2 = \lambda_3 = 200$. The initial state is assumed to be $x_0 = \{0.88, 0.0, 6.5\}$, and the initial value of the observer is $\hat{x} = \{0.8, 0.0, 5.0\}$. The simulation result for case 3 is shown in Figure 11.

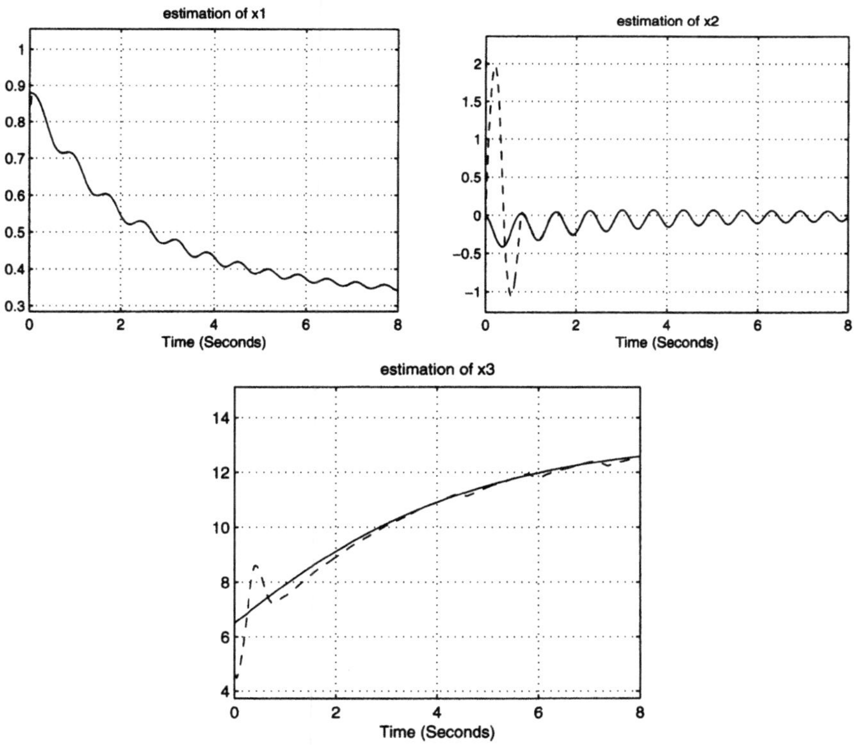

Fig. 11. Results of state estimation for a three-phase current motor using SMO

Example 5. To illustrate the fault diagnosis for a nonlinear system using SMFO, the three-phase current motor studied in the last example is considered. Its model is repeated as below:

$$\dot{x} = \begin{bmatrix} x_2 \\ -A_1x_2 - A_2x_3sinx_1 - A_3sin2x_1 \\ -D_1x_3 + D_2cosx_1 \end{bmatrix} + \begin{bmatrix} 0\,0 \\ 1\,0 \\ 0\,1 \end{bmatrix}\begin{bmatrix} u_1 \\ u_2 \end{bmatrix} + \begin{bmatrix} 0\,0 \\ 1\,0 \\ 0\,1 \end{bmatrix}\begin{bmatrix} f_{a1} \\ f_{a2} \end{bmatrix}$$

$$y = \begin{bmatrix} x_1 \\ x_3 \end{bmatrix}.$$

We consider the isolation of faults f_{a1} and f_{a2} through the multiple observer scheme. f_{a1} represents the fault of actuator 1 and the component faults which lead to change of parameters A_1, A_2 and A_3. f_{a2} represents the fault of actuator 2 and the component faults which lead to change of parameters D_1 and D_2.

First, we regard f_{a1} as an unknown input, and design SMFO as insensitive to f_{a1}. In this case, it is noted that x_3 is an unknown input free subsystem with one output. Because it is one-dimensional, the SMOFO design for x_3 is simplified as a simple SMOO. It is easy to show that x_1 and x_2 can be estimated using the SMFO technique proposed earlier, even if there is an unknown input in the x_d subsystem formed by x_1 and x_2. Generally, the estimation of x_d is required in order to form a SMOFO for x_o subsystem which is without unknown input. However, it is unnecessary here because x_3 is only affected by x_1, and x_1 is an output. Finally, the SMOFO is

$$\dot{z} = -D_1y_2 + D_2cos(y_1) + u_2 + \lambda sign(y_2 - z). \tag{99}$$

It is easy to show that this SMOFO is only sensitive to f_{a2}. The corresponding residual is

$$r_1 = (\lambda sign(y_2 - z)_{eq} = f_{a2}$$

In the second observer design, f_{a2} is considered as an unknown input. In this case, there is no unknown input free subsystem. However, the following SMO can be designed using the techniques discussed earlier

$$\begin{aligned} \dot{\hat{x}}_1 &= \hat{x}_2 + \lambda_1 sign(y_1 - \hat{x}_1) \\ \dot{\hat{x}}_2 &= -A_1\hat{x}_2 - A_2\hat{x}_3 siny_1 - A_3 sin2y_1 + u_1 + \lambda_2 sign((\lambda_1 sign(y_1 - \hat{x}_1)_{eq}) \\ \dot{\hat{x}}_3 &= -D_1\hat{x}_3 + D_2 cosy_1 + u_2 + \lambda_3 sign(\overline{e_3}) \end{aligned}$$

where

$$\overline{e_3} = \frac{(\lambda_2 sign(\overline{e_2}))_{eq}}{-A_2 siny_1} = e_3 + \frac{f_{a1}}{-A_2 siny_1}. \tag{100}$$

Assuming $\hat{f_{a1}} = \frac{f_{a1}}{-A_2 siny_1}$, we know that

$$r_2 = (\lambda_3 sign(\overline{e_3}))_{eq} = D_1\hat{f_{a1}} + f_{a2} + \dot{\hat{f}}_{a1}.$$

Therefore, we have the following fault diagnosis logic:

i. If $r_1 = r_2 \neq 0$, only f_{a2} happens.
ii. If $r_1 = 0, r_2 \neq 0$, only f_{a1} exists.
iii. If $r_1 \neq 0$, $r_2 \neq 0$ and $r_1 \neq r_2$, both faults f_{a1} and f_{a2} exist.

If $y_2 = x_3$ is not available, the SMOFO for fault diagnosis of actuator 2 cannot be designed, or r_1 will not be available. In this case, fault detection based on r_2 is possible, and fault isolation will be difficult. However, an abrupt fault of f_{a1} can still be isolated because only the abrupt change of f_{a1} will make the signal $\overline{e_3}$ (given by equation (100)) becomes nonzero for a short while. Therefore, we can record the third residual signal as $r_3 = \overline{e_3}$.

Figure 12 is the simulation result. For simplicity, it is assumed that the original estimation error is zero. The solid line and dashed line in subplot a) are the shapes of f_{a1} and f_{a2} respectively, and subplot b) is the residual signal r_1, c) is the residual signal r_2, d) is the residual signal r_3. Obviously, r_3 stays near zero for slow-varying f_{a1}, and only produces a nonzero value peak for abrupt f_{a1} at time $t = 18s$. The residuals r_1 and r_2 validate our fault isolation logic very well.

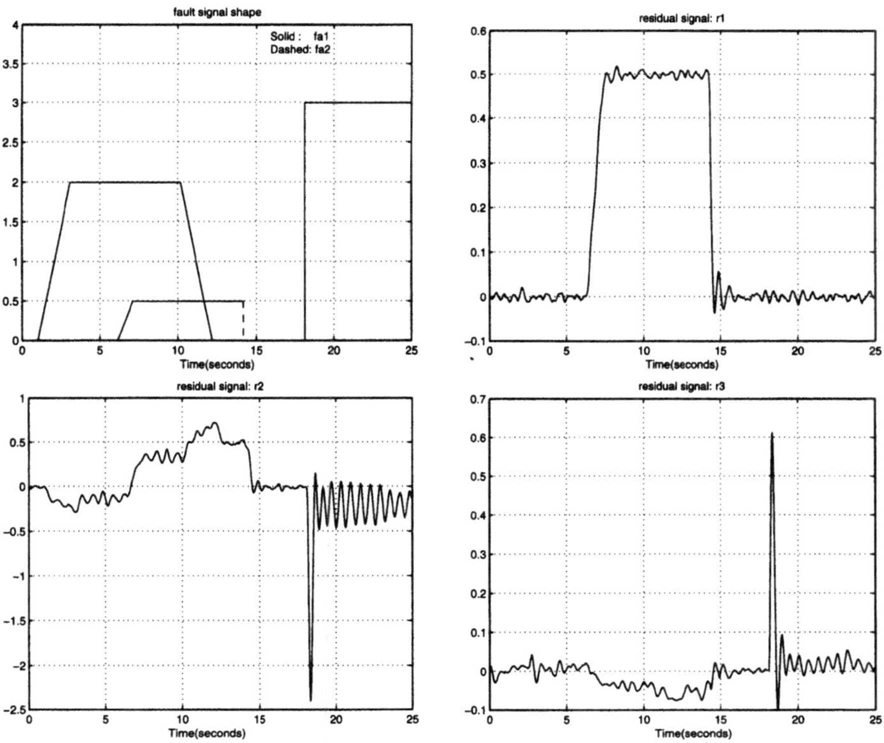

Fig. 12. Fault diagnosis simulation using SMFO for a nonlinear three phase motor model

9 Summary

In this chapter, we discuss the existence condition and design method for sliding mode observers and sliding mode functional observers which are insensitive to matched bounded uncertainty/nonlinearity. It is proved that the existence conditions of sliding mode observer are equivalent to those of classical unknown input observer. The sliding mode functional observer can exists under less conservative conditions compared with unknown input functional observer. The proposed design strategies were extended to a class of nonlinear systems. Additionally, the chapter explored application of the proposed SMO design strategies for fault detection and isolation in linear and nonlinear systems. A number of examples illustrated the application of the proposed methodologies to a number of mechatronics applications.

Acknowledgement

This research was sponsored in part by Natural Sciences and Engineering Research Council(NSERC) of Canada through its *Research* and *Strategic* grants program.

References

1. Barbot J.P, Boukhobza T., Djemai T.M. (1996) Sliding Mode Observer for Triangular Input Form. Proc. of 1996 IEEE Int. Conf. on Decision and Control, 1489–1490
2. Bestle D., Zeitz M. (1983) Canonical Form Observer Design for Nonlinear Time-Variable Systems. Int. Jour. Control 38:419-431
3. Birk J., Zeitz M. (1988) Extended Luenberger Observer for Nonlinear Multivariable Systems. Int. Jour. Control 47:1823–1836
4. Bornard G., Hammouri H. (1991) A High Gain Observer for a Class of Uniformly Observable Systems. Proc. of 1991 IEEE Int. Conf. on Decision and Control, Brighton
5. Boukhobza T., Djemai M., Barbot J.P. (1996) Nonlinear Sliding Observer for Systems in Output and Output Derivative Injection Form. Proc. of 13th IFAC World Congress, 299–304
6. Busawon K., Hammouri H., Bornard G. (1997) An Observer for a Class of Nonlinear Systems. Internal Report–LAGEP, University of Lyon I
7. Busawon K., Saif M. (1999) A State Observer for Nonlinear Systems. IEEE Trans. on Automatic Control 44:2098–2103
8. Chang S.K., Hsu P.L. (1995) A Novel Design for the Unknown Input Fault Detection Observer. Control Theory and Advanced Technology 10:1029–1052
9. Chen J., Patton R.J., Zhang H.Y. (1996) Design of Unknown Input Observers and Robust Fault Detection Filters. Int. Jour. of Control 63:85–105
10. Chung W. (1997) Game Theoretic and Decentralized Estimation for Fault Detection. Ph.D. Thesis, University of California, Los Angeles
11. Corless M., Tu J. (1998) State and Input Estimation for a Class of Uncertain Systems. Automatica 34:757–764

12. Djemai M., Barbot J.P., Bethoux O. (2000) On the Problem of Fault Detection and Residual Generation. Proc. of 2000 IEEE Int. Conf. on Decision and Control, 4335–4340
13. Douglas R. (1993) Robust Fault Detection Filter Design. Ph.D. Thesis, University of Texas at Austin
14. Drakunov S.V. (1992) Sliding Mode Observers Based on Equivalent Control Method. Proc. of 1992 IEEE Int. Conf. on Decision and Control, Tucson, Arizona, 2368–2369
15. Drakunov S.V., Utkin V. (1995) Sliding Mode Observers: A Tutorial. Proc. of 1995 IEEE Int. Conf. on Decision and Control, 3376–3378
16. Edwards C., Spurgeon S. (1994) On the Development of Discontinuous Observers. Int. Jour. Control 59:1211–1229
17. Frank P.M., Ding X. (1994) Frequency Domain Approach to Optimally Robust Residual Generation and Evaluation for Model-Based Fault Diagnosis. Automatica 30:789–804
18. Frank P.M., Wunnenberg J. (1989) Robust Fault Diagnosis Using Unknown Input Observer Schemes. In: Patton R.J. et al. (Eds.) Fault Diagnosis in Dynamic Systems: Theory and Application, 47–98
19. Frank P.M. (1994) Enhancement of Robustness in Observer-Based Fault Detection. Int. Jour. of Control 59:955–981
20. Garg V., Hedrick J.K. (1995) Fault Detection Filters for a Class of Nonlinear Systems. Proc. of 1995 American Control Conference, 1647–1651
21. Gauthier J.P., Hammouri H., Othman S. (1992) A Simple Observer for Nonlinear Systems – Application to Bioreactors. IEEE Trans. on Automatic Control 37(6)
22. Gauthier J.P., Kupka I.A.K. (1994) Observability and Observers for Nonlinear Systems. SIAM Jour. Control and Optimization 32:975–994
23. Guan Y., Saif M. (1991) A Novel Approach to the Design of Unknown Input Observers. IEEE Trans. on Automatic Control 36:632–635
24. Hammouri H., Kinnaert M., Yaagoubi E.H. (1998) Fault Detection and Isolation for State Affine Systems. European Jour. of Control 4:2–16
25. Hermans F.J.J., Zarrop M.B. (1996) Sliding Mode Observers for Robust Sensor Monitoring. Proc. of 13th IFAC World Congress, N:211–216
26. Horn R.A., Johnson C.A. (1985) Matrix Analysis, Cambridge University Press
27. Hou M., Busawon K., Saif M. (2000) Observer Design for Nonlinear Systems via Injective Mapping. IEEE Trans. on Automatic Control 45:1350-1355
28. Hou M., Muller P.C. (1994) Fault Detection and Isolation Observers. Int. Jour. of Control 60:827–846
29. Isidori A. (1995) Nonlinear Control Systems, 3rd Edition, Springer Verlag
30. Krener A.J., Isidori A. (1983) Linearization by Output Injection and Nonlinear Observers. Systems and Control Letters 3:47–52
31. Krener A.J., Respondek W. (1985) Nonlinear Observers with Linearizable Error Dynamics. SIAM Jour. Control and Optimization 23:197-216
32. Khalil H.K. (1996) Adaptive Output Feedback Control of Nonlinear Systems Represented by Input-Output Models. IEEE Trans. on Automatic Control 41:177–188
33. Kudva P., Viswanadham N., Ramarkrishna A. (1980) Observers for Linear Systems with Unknown Inputs. IEEE Trans. Automatic Control 25:113–115

34. Krishnaswami V., Rizzoni G. (1995) Model Based Health Monitoring of Vehicle Steering Systems Using Sliding Mode Observers. Proc. of 1995 American Control Conference, 1652–1656
35. Mangoubi R.S. (1995) Robust Estimation and Failure Detection for Linear Systems. Ph.D. Thesis, MIT, Cambridge
36. Massoumnia M.A. (1986) A Geometric Approach to Failure Detection and Identification in Linear Systems. Ph.D. Thesis, MIT, Cambridge
37. Misawa E.A. (1988) Nonlinear State Estimation Using Sliding Observers. Ph.D. Thesis, MIT
38. Patton R.J., Chen J. (1992) Robust Fault Detection of Jet Engine Sensor Systems Using Eigenstructure Assignment. Jour. of Guidance, Control and Dynamics 15:1491–1497
39. Raghavan S., Hedrick J.K. (1995) Observer Design for a Class of Nonlinear Systems. Int. Jour. Control 59:515–528
40. Rajamani R., Cho Y.M. (1998) Design of Observers for Nonlinear Systems. Int. Jour. of Control, 719–739
41. Rudolph J., Zeitz, M. (1994) A Block Triangular Nonlinear Observer Normal Form. Systems and Control Letters 23:1–8
42. Saberi A.P., Chen B.M., Sanutti P. (1993) Loop Transfer Recovery: Analysis and Design. Springer-Verlag, London
43. Saberi A.P., Chen B.M., Sanutti P. (1996) H_2 Optimal Control. Prentice Hall
44. Saif M. (1994) Towards a Model-Based Strategy to Automotive Engine Diagnostics. GM Restricted Tech. Report. Submitted to General Motors Research NAO R & D Center
45. Saif M., Guan Y. (1993) A New Approach to Robust Fault Detection and Identification. IEEE Trans. on Aerospace and Electronic Systems 29:685–695
46. Saif M. (1993) Reduced Order Proportional Integral Observer with Application. AIAA Jour. of Guidance, Control, and Dynamics 16:985–988
47. Saif M. (1994) A Disturbance Accommodating Estimator for Bilinear Systems. Control-Theory and Advanced Technology 10:431–446
48. Seliger R., Frank P.M. (1991) Fault Diagnosis by Disturbance Decoupled Nonlinear Observers. Proc. of 1991 IEEE Int. Conf. of Decision and Control, 2248–2253
49. Slotine J.-J., Hedrick J.K., Misawa E.A. (1987) On Sliding Observers for Nonlinear Systems. ASME Jour. of Dynamic Systems, Measurement and Control 109:245–252
50. Spong M. (1897) Modeling and Control of Elastic Joint Robots. ASME Jour. of Dynamic Systems, Measurement and Control 109:310–319
51. Steinberg A., Corless M.J. (1985) Output Feedback Stabilization of Uncertain Dynamical Systems. IEEE Trans. on Automatic Control 30:1025–1027
52. Sreedhar R., Fernandez B. (1993) Robust Fault Detection in Nonlinear Systems Using Sliding Mode Observers. IEEE Int. Conf. on Control Applications, 715–721
53. Tsui C.C. (1996) A New Design Approach to Unknown Input Observers. IEEE Trans. on Automatic Control 41:464–468
54. Utkin V. (1992) Sliding Modes in Control Optimization. Springer Verlag
55. Walcott B.L., Corless M.J., Zak S.H. (1987) Comparative Study of Nonlinear State-Observation Techniques. Int. Jour. of Control 45:2109–2132

56. Walcott B.L., Zak S.H. (1991) Combined Observer-Controller Synthesis for Uncertain Dynamical Systems with Application. IEEE Trans. on Systems, Man and Cybernetics 18:88–104
57. Wang H., Daley S. (1996) Actuator Fault Diagnosis: An Adaptive Observer-Based Technique. IEEE Trans. on Automatic Control 41:1073–1078
58. Wang G.-B., Peng S.S., Huang H.P. (1997) A Sliding Observer for Nonlinear Process Control. Chemical Engineering Science 52:787–805
59. Xia X.H., Gao W.B. (1989) Nonlinear Observer Design by Observer Error Linearization. SIAM Jour. of Control and Optimization 27:199–216
60. Xiong Y., Saif M. (1999) Functional Observers for Linear Systems with Unknown Inputs. Proc. of 14th IFAC World Congress, Bejing, China
61. Xiong Y., Saif M., (2000) Derivative Free State Functional and Unknown Input Estimation. Proc. of 2000 American Control Conference, Chicago
62. Xiong Y., Saif M. (2000) Robust Fault Detection and Isolation via Diagnostic Observer Design. Int. Jour. of Robust and Nonlinear Control–Special Issue on Fault Detection and Isolation 10:1175–1192
63. Xiong Y., Saif M. (2001) Sliding Mode Observer For Nonlinear Uncertain Systems. IEEE Trans. on Automatic Control 46:2012–2017
64. Yang H., Saif M. (1995) Fault Detection in a Class of Nonlinear Systems via Adaptive Sliding Observer Design. Proc. of 1995 IEEE Int. Conf. on Systems, Man and Cybernetics, Vancouver, BC, 2199–2204
65. Yang H., Saif M. (1996) Monitoring and Diagnostics of a Class of Nonlinear Systems Using a Nonlinear Unknown Input Observer. Proc. of 1996 IEEE Int. Conf. on Control Applications, 1006–1011
66. Yang H., Saif M. (1997) State Observation, Failure Detection and Isolation FDI in Bilinear Systems. Int. Jour. of Control 67:901–920
67. Yu D., Shields D.N., Daley S. (1996) A Bilinear Fault Detection Observer and its Application to a Hydraulic Drive System. Int. Jour. of Control 64:1023–1047
68. Yu D., Shields D.N. (1996) A Bilinear Fault Detection Observer. Automatica 32:1597–1602
69. Yu D. (1997) Fault Diagnosis for a Hydraulic Drive System Using a Parameter-Estimation Method. Control Engineering Practice 9:1283–1291
70. Zak S.H., Walcott B.L. (1990) State Observation of Nonlinear Control Systems via the Method of Lyapunov. Zinober A. (Ed.) Deterministic Control of Uncertain Systems, A.S.I., London

Fault Diagnosis and Fault Tolerant Control for Non-Gaussian Stochastic Systems with Random Parameters

Hong Wang

Department of Paper Science,
UMIST, Manchester M601QD, U. K.,
Institute of Automation, Chinese Academy of Sciences, P.R.China

Abstract. This chapter discusses the design of fault diagnosis and its related fault tolerant control for non-Gaussian stochastic systems subjected to parameter randomness. At first, a new formulation of fault diagnosis algorithm is proposed for linear fixed parameter systems that are subjected to non-Gaussian input. For this type of system, the residual signals are controlled, by the estimated fault, to reach a statistic state that is only affected by the original random inputs to the system and the uncontrollable part caused by the rate of changes of the unknown fault. This is followed by the design of fault diagnosis algorithm for non-Gaussian systems that are also subjected to random parameter changes. In this case, the fault is taken as the unexpected changes of the probability density functions of the random parameters. The Laplace transform is used to convey the output probability density function of the system into a simple form, where functional parameter estimation is applied to estimate the faults. Fault tolerant control has been formulated for both systems through an adaptive framework. A simulated example for the Thermal Mechanical Pulping process has been included to demonstrate the use of the proposed algorithm and interesting results have been obtained.

1 Introduction

In model based fault detection and diagnosis the faults are regarded as unexpected changes of some physical parameters of the system. The so far developed algorithms look into both deterministic and stochastic systems. The former uses observer based fault detection and diagnosis scheme [3,4,9–11], whilst the latter employs the Kalman filter techniques and on-line parameter identification [6]. In each cases, a residual signal can be generated and used to perform fault detection and diagnosis.

For stochastic systems, the residual signals are random processes whose statistic analysis can sometimes be difficult. However, many methods have been developed in the past two decades. In general, these approaches can be classified into the following two groups:

i. Identification based fault detection and diagnosis for dynamic systems whose models are unknown [6];
ii. Unexpected change detection for stochastic signals [7].

F. Caccavale, L. Villani (Eds.): Fault Diagnosis and Fault Tolerance for Mechatronic Systems, STAR 1, pp. 59-84, 2003.

The first approach uses an ARMAX model to represent the system and apply parameter identification, such as least square algorithms or stochastic gradient approaches, to estimate the unexpected changes in the system. As for the unexpected change detection for stochastic signals, focus has been largely made on either the detection of unexpected changes in the mean value and the variance of the considered random signals [7], or on the detection of unexpected parameter changes for static probability density functions of the random signal. In this case, the required fault detection and diagnosis are normally performed by applying the theory of statistical decision, where a likelihood ratio is evaluated between the hypotheses on the healthy parameters and the faulty parameters using the known probability density functions. As a result, the faulty parameters can be estimated by optimizing this likelihood ratio. These methods detect and diagnose the faults based on the residual generations but do not design the fault diagnosis scheme that directly minimizes the randomness of the residual signals.

In recent years, development has been made for a group of stochastic systems [15], where the system is represented in terms of the output probability density functions that are driven by a set of control input. Assuming that the output probability density functions are measurable, a B-spline decoupling model can be established, where the measured output probability density functions are represented by the instinct B-spline approximation whilst a set of differential equations are used to link the B-pline approximation weights to the control input. By expressing the relationship between the weights and the control input as a state space model, an observer based approach can be established that detects and diagnoses the fault in the system. In this case, the residual signal that is used to drive the observer is related to a weighted integration of the difference between the measured output probability density functions and the estimated output probability density functions.

However, all these methods are restricted to stochastic systems whose parameters are either fixed or obey a simple statistic distribution (e.g., Markovian jumping parameters). As such, there is a need to develop fault detection and diagnosis method that can be used to the stochastic systems that are subjected to arbitrary random parameters. These parameters are represented by their probability density functions whose unexpected changes are regarded as the fault occurring in the system. This type of system widely exist in process control practice and examples are the wet end control systems in paper manufacturing, particle size distribution control in food processing and chemical engineering, and combustion flames distributions. In fact, this type of system belongs to non-Gaussian stochastic systems, where standard mathematical treatment to fixed parameter systems cannot be directly applied.

The purpose of this chapter is therefore to develop fault detection and diagnosis algorithms for non-Gaussian systems that are subjected to random parameters. At first, a fixed parameter non-Gaussian linear system will be considered, where the fault detection and diagnosis algorithm will be formulated so that the residual signals only depend on the random input of the system. Indeed, if the fault detection and diagnosis can be performed by using the residual as a performance measure, then a clear interpretation of the statistics of the residuals in terms of stochastic disturbances will be obtained. That is, the estimated fault should be such that the residual signals have zero mean values and their statistics be affected only by the original random inputs to the system. It means that the optimal fault diagnosis scheme should be such that the randomness or uncertainty in the residual signals is minimized. Since the entropy is a general measure of the uncertainty of a random variable, the fault diagnosis can also be performed by minimizing the residual entropy. This is followed by the design for non-Gaussian stochastic systems that are subjected to random parameters. In this case, it is assumed that the output probability density functions are measurable. This condition enables the use of the Laplace transform to the probability density functions of the random parameters so as to transfer the complicated convolutions integration between the output probability density functions and those of random parameters into a simple algebraic equation. By applying the on-line estimation to this algebraic equation, the probability density function of the random parameters can be estimated. With the estimated fault, a fault tolerant control can be formulated by using an adaptive control framework.

The chapter is organized as follows: in sections 2 - 5, the fault detection and diagnosis algorithm for fixed parameter non-Gaussian stochastic systems will be described. Starting from section 6, the random parameter systems will be considered. This includes the fault diagnosis algorithm that minimizes the residual entropy in section 8 and the fault tolerant control in section 9. An applicability study will be given in section 10, which is then followed by concluding remarks.

2 The Representation for Fixed Parameter Systems

In the next few sections, we consider the following general linear and known stochastic system

$$x_{k+1} = Ax_k + Bfu_k + E\omega_k \tag{1}$$

$$y_k = Cx_k + Dfu_k + F\mu_k \tag{2}$$

where $x_k \in R^n$ is an unmeasurable state vector, $u_k \in R^m$ is the input vector, $y_k \in R^r$ is the measured output vector, respectively. $A \in R^{n\times n}$, $B \in R^{n\times m}$, $E \in R^{n\times p}$, $C \in R^{r\times n}$, $D \in R^{r\times m}$, and $F \in R^{r\times q}$ are known parameter matrices. $f \in R^{m\times m}$ represents the actuator gain of the system.

In equations (1) and (2), $\omega_k \in R^p$ and $\mu_k \in R^q$ are the two random processes representing either noises or model uncertainties. They can obey either Gaussian or non-Gaussian distributions. The problem to be solved is to use the measured input u_k and the output y_k to estimate unexpected changes in the actuator gain f. For this purpose, it is assumed that when the system is healthy [13],

$$f = f_H \in R^{m \times m} \tag{3}$$

where f_H stands for a healthy actuator gain and is assumed known. It is also assumed that the pair $\{A, C\}$ is observable.

3 Fault Detection for Fixed Parameter Systems

To perform the required fault detection, one can construct the following fault detection observer

$$z_{k+1} = Az_k + Bf_H u_k + L\bar{\epsilon}_k \tag{4}$$

$$\bar{\epsilon}_k = y_k - Cz_k - Df_H u_k \tag{5}$$

$$u_k \neq 0 \tag{6}$$

where $z_k \in R^n$ is the estimated state of the system, and $\bar{\epsilon}_k$ is the detection residual. $L \in R^{n \times r}$ is a fixed observer gain matrix which has been selected so that the matrix $A - LC$ is stable. This can be achieved since it has been assumed that $\{A, C\}$ is an observable pair. Note that the healthy actuator gain matrix, f_H, has been used here. Since all the matrices are known in equations (4)-(5), and the fault detection observer uses only the measured input and output of the system, this detection observer is therefore realizable in practice.

Define the following observation error

$$e_k = x_k - z_k \tag{7}$$

then the error dynamics can be formulated from equations (1) - (6) to give

$$\begin{aligned} e_{k+1} &= Ax_k + Bfu_k + E\omega_k - Az_k - Bf_H u_k - L\bar{\epsilon}_k \\ &= (A - LC)e_k + (B - LD)(f - f_H)u_k + \\ &\quad E\omega_k - LF\mu_k \end{aligned} \tag{8}$$

$$\bar{\epsilon}_k = Ce_k + D(f - f_H)u_k + F\mu_k \tag{9}$$

Denote

$$A_0 = A - LC \tag{10}$$

$$B_0 = B - LD \tag{11}$$

then it can be further obtained from equations (8)-(9) that

$$e_{k+1} = A_0 e_k + B_0(f - f_H)u_k + \eta_k \tag{12}$$
$$\bar{\epsilon}_k = Ce_k + D(f - f_H)u_k + \sigma_k \tag{13}$$
$$\eta_k = E\omega_k - LF\mu_k \tag{14}$$
$$\sigma_k = F\mu_k \tag{15}$$

where $\eta_k \in R^n$ and $\sigma_k \in R^n$ are the two random processes whose characteristics can be uniquely determined from those of ω_k and μ_k, respectively [8]. From equations (12) and (13), the transfer function of the detection residual, $\bar{\epsilon}_k$, can be readily formulated to give

$$\bar{\epsilon}_k = G(q^{-1})(f - f_H)u_k + v_k \tag{16}$$

where it has been denoted that

$$G(q^{-1}) = C(I - q^{-1}A_0)^{-1}q^{-1}B_0 + D \tag{17}$$
$$v_k = C(I - q^{-1}A_0)^{-1}q^{-1}\eta_k + \sigma_k \tag{18}$$

with v_k being another random process related to ω_k and μ_k. It can be seen that once the probability density functions of ω_k and μ_k are known, the probability density function $\gamma_v(Z)$ of v_k, can be directly calculated. From equation (16), it can be seen that when no fault occurs, i.e., $f = f_H$, then the residual becomes

$$\bar{\epsilon}_k = v_k \tag{19}$$

This means that, when the system is healthy, the residual is a random process which has the same known probability density function, $\gamma_v(Z)$, as that of v_k. This fact provides a fault detection mechanism which checks the statistics of the detection residual, $\bar{\epsilon}_k$. If the statistics are the same as that of v_k, then there is no fault in the system. Since v_k is calculated from two random inputs (i.e., ω_k and μ_k) that have known statistics, its mean value and variance, v_0 and σ_0, can be obtained from the statistics of ω_k and μ_k [8]. As such, by checking the changes in the mean values of $\bar{\epsilon}_k$, one can decide whether the system has a fault or not. Assume that at sample time k, there are a set of residual signals with window length N

$$\{\bar{\epsilon}_{k-N}, \bar{\epsilon}_{k-N+1}, \cdots, \bar{\epsilon}_k\} \tag{20}$$

then the following statistics on the mean and variance of $\bar{\epsilon}_k$ can be calculated to give

$$m_\epsilon(k, N) = \frac{1}{N+1} \sum_{i=k-N}^{k} \bar{\epsilon}_i \tag{21}$$

$$s_\epsilon(k, N) = \frac{1}{N} \sum_{i=k-N}^{k} (\bar{\epsilon}_i - m_\epsilon(k, N))^2 \tag{22}$$

As a result, the following parity check can be performed so as to detect faults in the actuator

$$if \left| \frac{m_\epsilon(k,N) - v_0}{s_\epsilon(k,N)/\sqrt{N+1}} \right| > \lambda \rightarrow f \neq f_H \tag{23}$$

where $\lambda > 0$ is a pre-specified threshold which reflects the confidence coefficient of the hypothesis test.

4 Fault Diagnosis for Fixed Parameter Systems

Once a fault is detected, its diagnosis needs to be performed so as to locate the size of the fault. This can be achieved by constructing an effective estimation algorithm for unknown $f \neq f_H$. Two cases, namely the constant actuator fault and a drifting fault, will be considered.

4.1 Fault Diagnosis for Constant Faults

In this subsection it is assumed that f is a constant and unknown matrix after a fault has occurred. To diagnose this type of fault, the following observer is used

$$\hat{x}_{k+1} = A\hat{x}_k + B\hat{f}_k u_k + L\epsilon_k \tag{24}$$

$$\epsilon_k = y_k - C\hat{x}_k - D\hat{f}_k u_k \tag{25}$$

where $\hat{x}_k \in R^n$ is the state, $\hat{f}_k$ is the estimate of the unknown f. All the other parameter matrices are the same as those defined in section 3. Defining the estimation error of fault diagnosis as

$$e_k = x_k - \hat{x}_k \tag{26}$$

and using the same formulation procedure as those in section 3, it can be shown that the residual signal ϵ_k satisfies

$$\epsilon_k = G(q^{-1})(f - \hat{f}_k)u_k + v_k \tag{27}$$

Since $f - \hat{f}_k$ is an estimation error, this residual consists of a linear combination of the estimation error and a random input that is only related to the original random inputs, ω_k and μ_k, to system (1). This suggests that the best estimation $\hat{f}_k$ should be such that the residual ϵ_k only relates to v_k. For this purpose, it is denoted that

$$G(q^{-1}) = \frac{M(q^{-1})}{N(q^{-1})} \tag{28}$$

$$M(q^{-1}) = \sum_{i=0}^{n} M_i q^{-i}, \quad M_i \in R^{r\times m} \tag{29}$$

$$N(q^{-1}) = 1 + \sum_{j=1}^{n} a_j q^{-j}, \quad a_j \in R^1 \tag{30}$$

where M_i and a_j are known and are directly related to the observer gain L and the parameter matrices of the original system (1) and (2). At the current sample time k, equation (27) can be formulated to give

$$\begin{aligned}\epsilon_k = G(q^{-1})(f - \hat{f}_{k-1} + \hat{f}_{k-1} - \hat{f}_k)u_k + \\ v_{k-1} + v_k - v_{k-1} = \epsilon_{k-1} \\ + G(q^{-1})(\hat{f}_{k-1} - \hat{f}_k)u_k + v_k - v_{k-1}\end{aligned} \tag{31}$$

where ϵ_{k-1} and $\hat{f}_{k-1}$ are available at sample time k. This equation enables us to design a recursive calculation of $\hat{f}_k$ so that ϵ_k is only affected by $v_k - v_{k-1}$, which is related to the original random inputs to the system. This means that $\hat{f}_k$ should be selected so that

$$\epsilon_{k-1} + G(q^{-1})(\hat{f}_{k-1} - \hat{f}_k)u_k = 0 \tag{32}$$

Using notations in equations (28) - (30), it can be seen from equation (32) that

$$N(q^{-1})\epsilon_{k-1} + M(q^{-1})(\hat{f}_{k-1} - \hat{f}_k)u_k = 0 \tag{33}$$

or in the time domain

$$\begin{aligned}\epsilon_{k-1} + a_1\epsilon_{k-2} + \cdots + a_n\epsilon_{k-n-1} + \\ M_0(\hat{f}_{k-1} - \hat{f}_k)u_k + M_1(\hat{f}_{k-2} - \hat{f}_{k-1})u_{k-1} + \cdots \\ + M_n(\hat{f}_{k-n-1} - \hat{f}_{k-n})u_{k-n} = 0\end{aligned} \tag{34}$$

Denote

$$\begin{aligned}\tilde{\epsilon}_k = \epsilon_{k-1} + a_1\epsilon_{k-2} + \cdots + a_n\epsilon_{k-n-1} + \\ M_1(\hat{f}_{k-2} - \hat{f}_{k-1})u_{k-1} + \cdots + \\ M_n(\hat{f}_{k-n-1} - \hat{f}_{k-n})u_{k-n}\end{aligned} \tag{35}$$

then it can be seen that at the current sample time k, $\tilde{\epsilon}_k$ is measurable. As a result, equation (34) becomes

$$\tilde{\epsilon}_k + M_0(\hat{f}_{k-1} - \hat{f}_k)u_k = 0 \tag{36}$$

Denote M_0^- as a pseudo inverse of matrix M_0 such that

$$M_0 M_0^- = I_r \tag{37}$$

and assume that $u_k \neq 0$, then by selecting the following adaptive tuning rule

$$\hat{f}_k = \hat{f}_{k-1} + M_0^- \frac{\tilde{\epsilon}_k u_k^T}{u_k^T u_k} \tag{38}$$

it can be shown that

$$(\hat{f}_{k-1} - \hat{f}_k) = -M_0^- \frac{\tilde{\epsilon}_k u_k^T}{u_k^T u_k} \tag{39}$$

As a result,

$$M_0(\hat{f}_{k-1} - \hat{f}_k)u_k = -M_0 M_0^- \frac{\tilde{\epsilon}_k u_k^T u_k}{u_k^T u_k}$$
$$= -\tilde{\epsilon}_k \tag{40}$$

This means that the adaptive tuning rule for $\hat{f}_k$ in equation (38) guarantees that equality (36) is satisfied so long as $u_k^T u_k \neq 0$. Using the adaptive tuning rule (38), it can be seen that the resulting residual satisfies

$$\epsilon_k = v_k - v_{k-1} \tag{41}$$

which has its mean value given by

$$\mathbf{E}\{\epsilon_k\} = 0 \tag{42}$$

4.2 Fault Diagnosis for Slow-Drifting Faults

The fault diagnosis algorithm developed in section 4.1 has considered the case of constant fault signals and the residual signal is only affected by the random inputs when the recursive fault diagnosis (38) is used. However, in practical systems, a drifting actuator fault can also occur where the gain of the actuator changes its value with respect to time. For such a system with a slow-changing fault, its state space model should be modified to

$$x_{k+1} = Ax_k + Bf_k u_k + E\omega_k \tag{43}$$
$$y_k = Cx_k + Df_k u_k + F\mu_k \tag{44}$$

When the same observer as that in (24)-(25) is used, the fault diagnosis residual is again given by

$$\epsilon_k = G(q^{-1})(f_k - \hat{f}_k)u_k + v_k \tag{45}$$

where $G(q^{-1})$ and v_k are the same as those in the previous sections. Similar to the procedures in the case of constant faults, equation (45) is re-formulated as

$$\begin{aligned}\epsilon_k &= v_{k-1} + v_k - v_{k-1} + \\ &\quad G(q^{-1})(f_k - f_{k-1} + f_{k-1} - \hat{f}_{k-1} + \hat{f}_{k-1} - \hat{f}_k)u_k \\ &= \epsilon_{k-1} + G(q^{-1})(\hat{f}_{k-1} - \hat{f}_k)u_k + v_k - v_{k-1} + \\ &\quad G(q^{-1})(f_k - f_{k-1})u_k\end{aligned} \tag{46}$$

Comparing with equation (31), the new term $G(q^{-1})(f_k - f_{k-1})u_k$ reflects the drifting actuator fault. From this structure of residual, it can be seen that the fault estimation should also be made to satisfy

$$\epsilon_{k-1} + G(q^{-1})(\hat{f}_{k-1} - \hat{f}_k)u_k = 0 \tag{47}$$

as the other three terms, v_k, v_{k-1} and $G(q^{-1})(f_k - f_{k-1})u_k$ in equation (46), cannot be minimized. Indeed, term $G(q^{-1})(f_k - f_{k-1})u_k$ is defined as an uncontrollable part that is purely caused by the unknown rate of change for the slow-drifting fault. Based upon this analysis, it can be seen that the same recursive diagnosis rule as equation (38) should still be used. In this context, the residual signal in equation (46) becomes

$$\epsilon_k = v_k - v_{k-1} + G(q^{-1})(f_k - f_{k-1})u_k \tag{48}$$

As a result, the residual ϵ_k is related to the random inputs and the control input weighted by the small changes of the faults between each time interval. This is a suboptimal solution and the mean value of this residual signal can be calculated from equation (48) to give

$$\mathbf{E}\{\epsilon_k\} = G(q^{-1})(f_k - f_{k-1})u_k \tag{49}$$

When the fault drifts at a slow rate, it can be assumed that there is a small and positive number ϵ_0 such that

$$\|f_k - f_{k-1}\| \leq \epsilon_0 \tag{50}$$

As such, when the input u_k is bounded by U_0 (i.e., $\|u_k\| \leq U_0$) and $G(q^{-1})$ is a stable polynomial, the following inequality can be established from equation (49)

$$\|\mathbf{E}\{\epsilon_k\}\| \leq G_0 \epsilon_0 U_0 \tag{51}$$

where G_0 is the l_1 norm of $G(q^{-1})$. From this equation it can be seen that for slowly drifting faults, so long as the magnitude of the control input is small, the mean value of the residual can also be made small.

5 Model Representation for Random Parameter Systems

Starting from this section, we consider the physical model that relates the input sequences u_k, the output sequence of system y_k and a stochastic noise term ω_k through the following ARMAX model

$$y_k = \sum_{i=1}^{n} a_i y_{k-i} + \sum_{j=1}^{m} b_j u_{k-j} + \omega_k \tag{52}$$

where $y_k \in R^1$ and $u_k \in R^1$ are one dimensional output and input of the system, respectively, and a_i, $(i = 1, 2, \cdots, n)$, b_j, $(j = 1, 2, \cdots, m)$, and ω_k are all independent and uniformly bounded random parameters characterized by their known probability density functions given by

$$P\{a \le y_k < \xi\} = \int_a^\xi \gamma_y(x, u_k)dx \tag{53}$$

$$P\{a \le a_i < \xi\} = \int_a^\xi \gamma_{ai}(x, \theta)dx \tag{54}$$

$$P\{a \le b_j < \xi\} = \int_a^\xi \gamma_{bj}(x, \theta)dx \tag{55}$$

$$P\{a \le \omega_k < \xi\} = \int_a^\xi \gamma_\omega(x)dx \tag{56}$$

where $\theta \in R^L$ is a parameter vector whose unexpected changes are regarded as the faults in the system. In equation (52), n and m are known structure orders of the system. Without loss of generality, we assume that all the parameters are positive. This means that $a > 0$ for all the probability density functions. We also assume that:

A1 the triple $\{\gamma_y(x, u_k), y_k, u_k\}$ are measurable;

A2 Under healthy conditions, $\theta = \theta_H$ and $\{\gamma_{ai}(x, \theta_H), \gamma_{bj}(x, \theta_H)\}$ are known.

The purpose of fault detection and diagnosis is therefore to use the triple $\{\gamma_y(x, u_k), y_k, u_k\}$ to detect unexpected changes in parameter vector θ.

Suppose that system (52) is stable under the conditions stated in the theorem of Kharitonov, and u_k is bounded by M_u, then the output sequence y_k of equation (52) is also a bounded stochastic process at sample time k. This means that the probability density function of y_k can be defined on the bounded interval $[a, b]$ and is of course related to the past inputs and outputs given by

$$\phi_k = \{y_{k-1}, y_{k-2}, ..., y_{k-n}, u_k, u_{k-1}, ..., u_{k-m}\} \tag{57}$$

Since at sample time k, ϕ_k is measurable, the output probability density function at sample time k is in fact a conditional probability density function under available ϕ_k. To detect and diagnose the fault, it is important to formulate a mathematical relationship that links the measured $\gamma_y(x, u_k)$ with $\{\gamma_{ai}(x, \theta_H), \gamma_{bj}(x, \theta_H)\}$ for all $i = 1, 2, ..., n$ and $j = 1, 2, ..., m$.

Since ϕ_k is measurable at sample time k, it can be seen from equation (52) that y_k is a linear combinations of all the random parameters and the random noise, ω_k. Under the assumption that all the random parameters and the random noise are independent, then from the results in the probability theory, it can be shown that the probability density function $\gamma(x, u_k)$ is of a

$(n+m+1)$-fold convolution of the form

$$\gamma_y(x,u_k) =$$
$$\int_a^x\int_a^b\cdots\int_a^b \gamma_\omega(x-z_{n+m})\prod_{j=0}^{m-1}\gamma_{b(m-j)}(z_{n+m-j}-z_{n+m-j-1},\theta,u_{k-m+j})\times$$
$$\times\prod_{i=0}^{n-2}\gamma_{a(n-i)}(z_{n-i}-z_{n-i-1},\theta,y_{k-n+i})\gamma_{a1}(z_1,\theta,y_{k-1})dz_1dz_2\cdots dz_{n+m} \quad (58)$$

where $\gamma_{ai}(x,\theta,y_{k-i})$ is the probability density function of random variable a_iy_{k-i}, and $\gamma_{bj}(x,\theta,u_{k-j})$ is the probability density function of random variable b_ju_{k-j}, respectively.

Denote $\hat{\theta}(k)$ as the estimate of θ, then the estimated output probability density function can be expressed as

$$\gamma_y(x,u_k,\hat{\theta}) =$$
$$\int_a^x\int_a^b\cdots\int_a^b \gamma_\omega(x-z_{n+m})\prod_{j=0}^{m-1}\gamma_{b(m-j)}(z_{n+m-j}-z_{n+m-j-1},\hat{\theta}(k),u_{k-m+j})\times$$
$$\times\prod_{i=0}^{n-2}\gamma_{a(n-i)}(z_{n-i}-z_{n-i-1},\hat{\theta}(k),y_{k-n+i})\gamma_{a1}(z_1,y_{k-1})dz_1dz_2\cdots dz_{n+m} \quad (59)$$

Since $\gamma_y(x,u_k)$ is measurable, one can select $\hat{\theta}(k)$ such that the following performance function

$$J=\int_a^b(\gamma_y(x,u_k)-\gamma_y(x,u_k,\hat{\theta}))^2dx=\min \quad (60)$$

As such, the fault detection and diagnosis can be performed by a nonlinear minimization to J which measures the functional distance for the functional space that contains all the continuous functions defined on $[a,b]$. However, since there are $(n+m+1)$-multiple integrations involved in calculating J, it is generally difficult to implement such a complicated optimization process.

This leads to the re-consideration of the expression for the output probability density function as shown in equation (58). The idea comes from the Laplace transforms and the transfer functions in classical control systems theory, where it has been shown that in terms of the Laplace transformed variables, the convolutions in time domain variables can be transferred as an algebraic multiplication in s-domain. As such, if we can apply the Laplace transforms to all the probability density functions of the random parameters and the noise, the $(n+m+1)$-fold convolutions in equation (58) can then be expressed as an algebraic multiplications of the transformed probability density functions. This is a new approach that has recently been developed in Ref. [17]. In the next section, the s-domain expression of the output probability density function will be described.

6 Laplace Transformations for Probability Density Functions

In this section, we consider the Laplace transformations for general probability density functions that are defined on $[0, +\infty)$ interval. Denote such a probability density function as $\gamma(x)$ with $x \in [0, +\infty)$, then its Laplace transform is defined as

$$\Gamma(s) = \int_0^{+\infty} \gamma(x) e^{-sx} dx \tag{61}$$

where s is the Laplace variable. For this definition, the following properties can be directly established.

Property 1: Since all the probability density functions are non-negative, we have $\Gamma(s) \geq 0$ for all $s > 0$;

Property 2: The mean value of the random variable x that has the probability density function of $\gamma(x)$ is given by

$$m_x = E\{x\} = \int_0^{+\infty} x\gamma(x) dx = -\frac{d\Gamma(s)}{ds}|_{s=0} \tag{62}$$

Property 3: The variance of random variable x that has a probability density function $\gamma(x)$ is given by

$$E\{(x - m_x)^2\} = E\{x^2\} - m_x^2 = \frac{d^2\Gamma(s)}{ds^2}|_{s=0} - (\frac{d\Gamma}{ds}|_{s=0})^2 \tag{63}$$

It can also be formulated that for the lth momentum, we have

$$E\{x^l\} = \int_0^{+\infty} x^l \gamma(x) dx = (-1)^l \frac{d^l\Gamma(s)}{ds^l}|_{s=0}$$

From these properties, it can be seen that all the distribution properties of a probability density functions can be calculated using the Laplace transformed functions. Denote the Laplace transforms of the probability density functions of $a_i y_{k-i}$ and $b_j u_{k-j}$ as $\Gamma_{ai}(s, \theta, y_{k-i})$ and $\Gamma_{bj}(s, \theta, u_{k-j})$ respectively, then by applying the Laplace transforms to equation (58), it can be shown that

$$\Gamma_y(s, u_k) = \Gamma_\omega(s) \prod_{j=1}^{m} \Gamma_{bj}(s, \theta, u_{k-j}) \prod_{i=1}^{n} \Gamma_{ai}(s, \theta, y_{k-i}) \tag{64}$$

where

$$\Gamma_y(s, u_k) = \int_0^{+\infty} \gamma_y(x, u_k) e^{-sx} dx \tag{65}$$

$$\Gamma_{ai}(s,\theta,y_{k-i}) = \int_0^{+\infty} \gamma_{ai}(x,\theta,y_{k-i})e^{-sx}dx \tag{66}$$

$$\Gamma_{bj}(s,\theta,u_{k-j}) = \int_0^{+\infty} \gamma_{bj}(x,\theta,u_{k-j})e^{-sx}dx \tag{67}$$

Since in the probability theory there are the following relationships between $\gamma_{ai}(x,\theta)$, $\gamma_{ai}(x,\theta,y_{k-i})$, $\gamma_{bj}(x,\theta)$ and $\gamma_{bj}(x,\theta,u_{k-j})$ as

$$\gamma_{ai}(x,\theta,y_{k-i}) = \frac{1}{|y_{k-i}|}\gamma_{ai}(\frac{x}{|y_{k-i}|},\theta) \tag{68}$$

$$\gamma_{bj}(x,\theta,u_{k-j}) = \frac{1}{|u_{k-j}|}\gamma_{tj}(\frac{x}{|u_{k-j}|},\theta) \tag{69}$$

we have, in terms of the Laplace transforms, that

$$\Gamma_{ai}(s,\theta,y_{k-i}) = \begin{cases} \Gamma_{ai}(y_{k-i}s,\theta), & y_{k-i} > 0 \\ -\Gamma_{ai}(y_{k-i}s,\theta), & y_{k-i} < 0 \\ 1, & y_{k-i} = 0 \end{cases} \tag{70}$$

and

$$\Gamma_{ai}(s,\theta,y_{k-i}) = \begin{cases} \Gamma_{bj}(u_{k-j}s,\theta), & u_{k-j} > 0 \\ -\Gamma_{bj}(u_{k-j}s,\theta), & u_{k-j} < 0 \\ 1, & u_{k-j} = 0 \end{cases} \tag{71}$$

As a result, equation (64) can be further expressed, in terms of the Laplace transforms of the probability density functions of random parameters (a_i, b_j), as

$$\Gamma_y(s,u_k) = \Gamma_\omega(s)\prod_{j=1}^{m}\Gamma_{bj}(u_{k-j}s,\theta)\prod_{i=1}^{n}\Gamma_{ai}(y_{k-i}s,\theta) \tag{72}$$

where, without loss of generality, it has been assumed that all the input and output measurements are positive. Denote $\Gamma_y(s,\hat{\theta}(k),u_k)$ as the Laplace transform of $\gamma_y(x,\hat{\theta}(k),u_k)$, then convolutions (59) can be similarly expressed as

$$\Gamma_y(s,u_k,\hat{\theta}(k)) = \Gamma_\omega(s)\prod_{j=1}^{m}\Gamma_{bj}(u_{k-j}s,\hat{\theta}(k))\prod_{i=1}^{n}\Gamma_{ai}(y_{k-i}s,\hat{\theta}(k)) \tag{73}$$

To further reduce the calculation of the multiplication, we introduce the following performance function in the s-domain for the estimation

$$\pi(\hat{\theta}(k)) = \int_0^{+\infty} K(s)(log\frac{\Gamma_y(s,\hat{\theta}(k),u_k)}{\Gamma_y(s,u_k)})^2 ds \tag{74}$$

where $K(s)$ is a weighting function to be selected so that for any bounded $\{\Gamma_y(s,u_k), \Gamma_y(s,\hat{\theta}(k),u_k)\}$, the performance function should satisfy

$$\left|\pi(\hat{\theta}(k))\right| < +\infty \tag{75}$$

In a recent work [17], the following weight function

$$0 \leq K(s) \leq \frac{M_1 s^2 e^{-M_2 s}}{8M_0^2\left[(logM_0)^2 + (logs)^2\right]} \tag{76}$$

has been selected, where M_0 is the uniform upper bound of the output probability density function (i.e., $|\gamma_y(x,u_k,\theta)| \leq M_0$), and M_1 and M_2 are prespecified positive numbers.

The advantage of using "*log*" operation is to transfer the multiplications in equations (72) and (73) into algebraic sum. Since the output probability density function $\gamma_y(x,u_k)$ is assumed measurable, this performance, as indicated already, is only a function of the estimated parameter $\hat{\theta}(k)$. Indeed, because there is a one-to-one relationship between the probability density function of the random variable and its Laplace transform, to estimate the unexpected changes in the probability density function of the random parameters, we only need to estimate the unexpected changes in $\Gamma_{ai}(y_{k-i}s,\theta)$ and $\gamma_{bj}(u_{k-j}s,\theta)$. In this context, the estimation of θ can be performed by minimizing $\pi(\hat{\theta}(k))$ as defined in equation (74). This is a nonlinear optimization problem, where an analytic solution does not exist in general. Of course, one can use the following gradient rule to get a "kind" of solution as

$$\hat{\theta}(k) = \hat{\theta}(k-1) - \lambda\frac{\partial\pi}{\partial\theta}|_{\theta=\hat{\theta}(k-1)} \tag{77}$$

In the next section, the scanning parameter estimation method described in [14] will be used.

7 Unexpected Change Diagnosis Using Scanning Parameter Estimation

Again, by applying "*log*" operation to both sides of equation (73), it can be obtained that

$$log(\Gamma_y(s,u_k,\hat{\theta}(k))) = \tag{78}$$

$$log(\Gamma_\omega(s)) + \sum_{j=0}^{m} log(\Gamma_{bj}(u_{k-j}s,\hat{\theta}(k))) + \sum_{i=1}^{n} log(\Gamma_{ai}(y_{k-i}s,\hat{\theta}(k)))$$

At this stage, since the output probability density function $\gamma_y(x,u_k,\theta)$ and $\gamma_\omega(x)$ are known, by denoting

$$r(s,k) = log(\Gamma_y(s,u_k,\hat{\theta}(k))) - log(\Gamma_\omega(s)) \tag{79}$$

then $r(s,k)$ is measurable. This transforms equation (78) into

$$r(s,k) = \sum_{j=0}^{m} log(\Gamma_{bj}(u_{k-j}s, \hat{\theta}(k))) + \sum_{i=1}^{n} log(\Gamma_{ai}(y_{k-i}s, \hat{\theta}(k))) \tag{80}$$

Assuming that there are a set of basis functions $B_p(s)$ with $p = 1, 2, ..., N$ so that each function in equation (80) can be approximated as

$$log(\Gamma_{bj}(u_{k-j}s, \hat{\theta}(k)) = \sum_{p=1}^{N} v_p^{bj}(\hat{\theta}(k))B_p(u_{k-j}s) \tag{81}$$

$$log(\Gamma_{ai}(y_{k-i}s, \hat{\theta}(k)) = \sum_{p=1}^{N} v_p^{ai}(\hat{\theta}(k))B_p(y_{k-i}s) \tag{82}$$

$$\tag{83}$$

we can then select a set of $s_q \in [0, +\infty)$ for $q = 1, 2, ..., M$ such that

$$r(s_q, k) = \sum_{j=0}^{m}\sum_{p=1}^{N} v_p^{bj}(\hat{\theta}(k))B_p(u_{k-j}s_q) + \sum_{i=1}^{n}\sum_{p=1}^{N} v_p^{ai}(\hat{\theta}(k))B_p(y_{k-i}s_q) \tag{84}$$

where the weights in front of the basis functions are only functions of $\hat{\theta}(k)$. Using this decomposition expression, we can then express equation (80) in the following format

$$r(s_q, k) = \Theta(k)\Phi(k, s_q) \tag{85}$$

$$\begin{aligned}\Theta(k) = (&v_1^{b1}, v_2^{b1}, ..., v_N^{b1} v_1^{b2}, v_2^{b2} ..., v_N^{b2}, \\ &......, v_1^{a1}, v_2^{a1}, ..., v_N^{a1}, ..., \\ &v_1^{an}, v_2^{an}, ..., v_N^{an})\end{aligned} \tag{86}$$

$$\begin{aligned}\Phi(k, s_q) = (&B_1(u_k s_q), B_2(u_k s_q), ..., B_N(u_k s_q), \\ &B_1(u_{k-1} s_q), B_2(u_{k-1} s_q), ..., B_N(u_{k-1} s_q), ..., \\ &..., B_1(y_{k-n} s_q), B_2(y_{k-n} s_q), ..., B_N(y_{k-n} s_q))\end{aligned} \tag{87}$$

Then in terms of index $q = 1, 2, ..., M$, the following recursive least square algorithm can be applied

$$\Theta^q(k) = \Theta^{q-1}(k) + \frac{P(q)\Phi(k, s_q)\epsilon_q(k, s_q)}{1 + \Phi(k, s_q)^T P(q)\Phi(k, s_q)} \tag{88}$$

$$\epsilon_q(k, s_q) = r(s_q, k) - \Theta^{q-1}(k)\Phi(k, s_q) \tag{89}$$

$$P^{-1}(q) = P^{-1}(q-1) + \Phi(k, s_q)\Phi(k.s_q)^T \tag{90}$$

The estimated vector $\Theta(k)$ should therefore be given by

$$\Theta(k+1) = \Theta^M(k) \tag{91}$$

Using this recursive calculation, the unexpected change detection can be performed by the following mechanism

$$|\epsilon_q(k, s_q)| > \lambda \rightarrow a \;\; fault \;\; has \;\; occurred \tag{92}$$

8 Fault Diagnosis Design via Minimizing Residual Entropy

So far we have discussed two approaches for the fault detection and diagnosis of stochastic systems, where the unexpected changes are detected and diagnosed through the residual signals as in equations (23), (38) and (92). The idea is to minimize residual uncertainty for linear stochastic systems with either fixed or random parameters. In this section, we continue this idea by considering general nonlinear stochastic dynamic systems which are subjected to explicit random input

$$y_k = h(y_{k-1}, y_{k-2}, \cdots, y_{k-n}, u_k, \cdots, u_{k-m}, F, \omega_k) \tag{93}$$

where ω_k is an known arbitrary bounded random input and F is the fault and $h(...)$ is a general nonlinear function that characterizes the dynamics of the nonlinear system. For such a system, assuming $h(...)$ is known, then the following estimated model

$$\hat{y}_k = h(y_{k-1}, y_{k-2}, \cdots, y_{k-n}, u_k, \cdots, u_{k-m}, \hat{F}, \omega_k) \tag{94}$$

can be obtained, leading to the following simple residual signal

$$\epsilon(k) = \hat{y}_k - y_k \tag{95}$$

Denote

$$\phi(k) = [y_{k-1}, \cdots, y_{k-n}, u_k, \cdots, u_{k-m}] \tag{96}$$

then at sample time k, $\phi(k)$ is available and the residual signal can be further expressed as

$$\epsilon(k) = g(\phi(k), \hat{F}, \omega_k) \tag{97}$$

where $g(.,.,.)$ is another known nonlinear function which is related to the original system function $h(...)$. Assuming that

- the probability density function of the noise is $\gamma_\omega(x)$, and
- function $g(.,.,.)$ in invertible with respect to variable term ω_k

then at sample k, the residual signal has a probability density function $\gamma_\epsilon(x, \phi(k), \hat{F})$, that is given by

$$\gamma_\epsilon(x, \phi(k), \hat{F}) = \gamma_\omega(g^{-1}(\phi(k), \hat{F}, x)) \left| \frac{\partial g^{-1}(\phi(k), \hat{F}, x))}{\partial x} \right| \tag{98}$$

In this case, the best estimation $\hat{F}$ at sample instance k should be solved by minimizing the mean values of the residual and its entropy (i.e., the uncertainty)

$$J(\hat{F}) = -\int_a^b \gamma_\epsilon(x, \phi(k), \hat{F}) log(\gamma_\epsilon(x, \phi(k), \hat{F}))dx + \int_a^b x\gamma_\epsilon(x, \phi(k), \hat{F})dx \tag{99}$$

where the first term is the residual entropy. This is simply because that a good fault estimation should have a very small mean value of its residual, and at the same time the uncertainty of the residual signal should be made as small as possible. Since the entropy is a more general measure than just variance of uncertainty (randomness) for non-Gaussian distributions, it is used here for the selection of the estimated fault. To minimize $J(\hat{F})$, one can simply use the following gradient learning rule

$$\hat{F}(k) = \hat{F}(k-1) - \lambda \frac{\partial J}{\partial F}|_{F=\hat{F}(k-1)} \tag{100}$$

where $\lambda > 0$ is a pre-specified learning rate. Similar formulations can also be obtained for continuous-time systems represented by the following Ito differential equation

$$\dot{x} = f(x, u, F)dt + \sigma(x, u, F)dw \tag{101}$$

where x is a state vector and w is a stochastic process.

9 Fault Tolerant Control Design

Once the fault is diagnosed, controller that is designed for the healthy system should be modified so as to compensate the performance losses that are caused by the fault in the system. In this respect, there are only two situations as listed below:

- the fault has caused a major performance deterioration in the closed loop system, and
- the fault has caused certain performance losses in the closed loop system.

In the first case, the solution to the problem would generally be to switch off the system and perform the required repair. For example, when all the actuators fail to operate, there is no need to modify the controller at all and in this case the use of fault tolerant control is not suited. However, in the second case, it is generally necessary to modify the control input so that the closed loop system can still be operated. This needs to the design of fault tolerant control. This is because if only some part of the actuators failed, one can still use other actuators to perform the control task. Of course, this requires the controller to be redesigned so as to realize the closed loop control through the remaining healthy actuators. This can be achieved via the use of either the reconfiguration of the system or the use of an adaptive control framework. In this respect, the well developed adaptive control can be regarded as the fault tolerant control.

In this section, both the direct control modification and the adaptive control will be formulated for the stochastic systems discussed so far in this chapter.

9.1 Fault Tolerant Control for Fixed Parameter Systems

In this context, we consider again the system represented by equations (1) and (2). We assume that the original healthy system has an output based controller of the following format:

$$u_k = K(y_k, f_H, r_k) \tag{102}$$

where K is a control function that realizes the closed loop control for the healthy system and $r_k \in R^r$ is the set point for the closed loop system. It can be concluded that is closed loop control includes the value of the healthy actuator matrix f_H.
Without of loss of generality, we can assume that this controller stabilizes the closed loop system when $r_k = 0$ and there is an upper-bound η so that x_k

$$\|x_k\| \leq \eta \tag{103}$$

for any bounded $E\omega_k$ and $F\mu_k$. The fault tolerant control design looks into two aspects. In the first aspect, we assume that the fault is not serious in the sense that $rank(f) = m$. In this case, an adaptive control scheme can be used for the control input (102) to give

$$u_k = \hat{f}_k^{-1} f_H K(y_k, \hat{f}_k, r_k) \tag{104}$$

where $\hat{f}_k$ is obtained from equation (38). In this case the control input can be formulated further as

$$u_k = \hat{f}_k^{-1} f_H K(y_k, f_H, r_k) + u_0 \tag{105}$$

$$u_0 = \hat{f}_k^{-1} f_H \frac{\partial K}{\partial f}|_k(\hat{f}_k - f_H) \tag{106}$$

Assuming the fault diagnosis has realized a good estimation in the sense that $\hat{f}_k$ is close to f, then when control input (105)-(106) is applied to the system, the closed loop equation becomes

$$x_{k+1} = Ax_k + Bf_H K(y_k, f_H, r_k) + Bf_H \frac{\partial K}{\partial f}|_k(\hat{f}_k - f_H)u_k + E\omega_k \tag{107}$$

$$y_k = Cx_k + Df_H K(y_k, f_H, r_k) + Df_H \frac{\partial K}{\partial f}|_k(\hat{f}_k - f_H)u_k + F\mu_k \tag{108}$$

From this equation, it can be seen that the stability of the closed loop system can still be guaranteed if the following condition is satisfied

$$\|f_H \frac{\partial K}{\partial f}|_k(\hat{f}_k - f_H)u_k\| \leq +\infty \tag{109}$$

When some of the actuators are totally failed, the gain matrix f will be singular. In this case, the reconfiguration of the controller is required, where the system equation can be re-written in the form of a reduced actuators form. In this context, the fault diagnosis (38) is only used to help such a reconfiguration. A different control has to be verified. The reconfigured system should be of the following form

$$x_{k+1} = Ax_k + B_r f_r u_k^r + E\omega_k \tag{110}$$

$$y_k = Cx_k + D_r f_r u_k^r + F\mu_k \tag{111}$$

where $B_r f_r u_k^r$ and $D_r f_r u_k^r$ are the reconfigured input channel for the system and u_k^r is the reduced dimensional input to the system that applies only to the remaining healthy actuators. This requires that the control input given by equation (102) be able to be reconfigurable in terms of the diagnosed faults $\hat{f}_k$. A switching function has therefore to be built into the controller based on the values of $\hat{f}_k$. For example, if some values of the elements in matrix $\hat{f}_k$ are very small, then these actuators should be switched off. Since the system is linear, the well developed linear control or adaptive control can be applied to find out the control function for the reduced actuators system (110)-(111).

9.2 Fault Tolerant Control for Random Parameter Systems

The fault tolerant control for system (52) can be obtained by purely using an adaptive framework. In this case the design purpose is to obtain a control sequence so that the shape of the output probability density function $\gamma_y(x, u_k)$ is made as close as possible to a given distribution function. This controller should have an adaptive component for the estimated probability density functions of the random parameters as generated by equations (88) - (90). Denote such a desired distribution function as $g(x)$, then its Laplace transform can be obtained to give

$$\Gamma(s) = \int_0^{+\infty} g(x)e^{-sx}dx \tag{112}$$

Using the same performance function as given by Wang et, al, (2002), it can be obtained that

$$J = \int_0^{+\infty} K(s)(log\frac{\Gamma_y(s, \hat{\theta}(k), u_k)}{\Gamma(s)})^2 ds + Ru_k^2 \tag{113}$$

where $R > 0$ is a pre-specified weighting function. Using this performance function, the control design for known probability density functions of the random parameters have been carried out in Wang (2002). This will form a basis for the adaptive control by simply incorporating the estimation of the probability density functions in (88)-(90). Detailed design procedure will therefore not be given here.

10 Applicability Study

In this section, we consider a simple example in the refining system in paper making process. Refining is a key part in paper making in generating fibres from raw materials such as wood and straw. This is an energy intensive process that consumes up to 75 percent of total energy used by paper mills. For example in Thermo-Mechanical Pulping (TMP), up to 250 kWhr/tonne of electrical energy can be consumed. As such, effective and reliable control in this part of the paper making will therefore

- increase the produced fibre quality;
- reduce energy consumption, and
- minimizing the faults occurring in the whole process.

10.1 Process Description

As described in [16], a chip refiner consists of either two counter-rotating or one-fixed and one rotating grooved plates with pressure exerted on one of them by a hydraulic cylinder. This structure of such a chip refiner is shown in Fig. 1, where wood chips and diluition water are fed near the axis and are forced to move outward between the plates by centrifugal and frictional forces between the two plates. The wood chip feeding is realized via a transfer

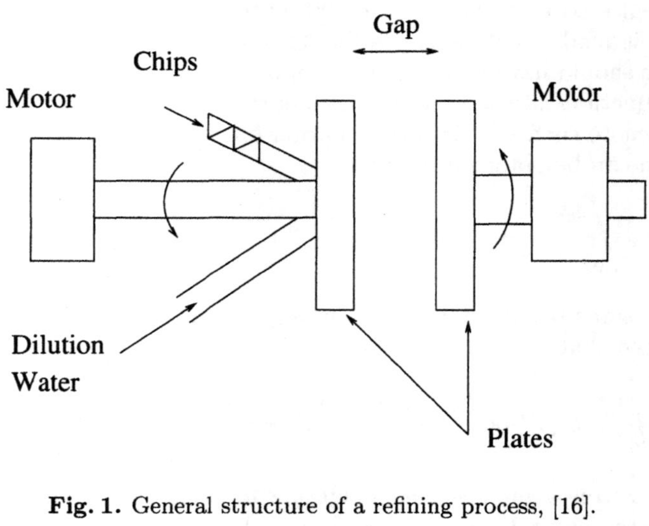

Fig. 1. General structure of a refining process, [16].

screw shown in the figure and the two plates are driven by electrical motors to a constant high speed. In some cases, only one plate rotates and the other is fixed.

Due to mechanical actions and high pressure steam and friction, chips are broken down into fibres, which, when combined with steam produced by evaporation, creates a few hundred micrometer thick pad between the plates. This pad forms a major load to the driving motor [1,2]. This is a continuous process where wood chips are fed in continuously and the pulp and steam are generated and discharged into the next stage.

To characterize the quality of the produced pulp, a number called freeness F is used which describes the resistance of a fiber network to the flow of water. A widely applied approach to determine the freeness is to use a Canadian Standard Freeness (CSF) tester discussed in [12]. However, since the freeness measurement is only taken every one or two hours in most paper mills, such a long sampling period will not be suitable for on-line control purposes. Instead, another quantity called specific energy is used. This means that there is a deterministic function, η which relates freeness and specific energy S_e

$$F = \eta(S_e) \tag{114}$$

in the TMP refining process. The specific energy is defined as energy consumed per mass unit of wood fibres and is expressed as

$$Specific\ Energy = \frac{Motor\ Load}{Oven\ Dry\ Pulp\ Mass\ Flow} \tag{115}$$

The purpose of closed loop control for refining process is to manipulate either the wood chip feed $g(k)$ or the gap in order to control the process to achieve the desired motor load, specific energy and consequently, the freeness of the fibre.

To control the motor load, the wood chip feeding is manipulated by adjusting the transfer screw. In general, the increased wood feeding speed will increase the motor load and thus increase the specific energy. This character is shown in Fig. 2. Assuming the current operation of the refiner is at a wood chip feeding speed g_0 that produces a steady motor load M_0, then the following normalized incremental values for the wood chip feeding and motor load are used to establish the model.

$$u_k = \frac{g(k) - g_0}{g_0}$$

$$y_k = \frac{M(k) - M_0}{M_0}$$

where $g(k)$ and $M(k)$ are the absolute values of the wood chip feeding speed and motor load at sample time k, respectively. By taking the input to the system as u_k and the output from the system as y_k, a dynamic nonlinear system of the form (1)-(2) can be established around an operating point in the characteristic curve shown in Fig. 2. In general, a second order dynamic system with fixed parameters can be obtained. This differs from the control system where the plate gap is used as an input, as in this case the dynamic equation is of a nonlinear nature where the gain of the control input channel is dependent of the current gap position.

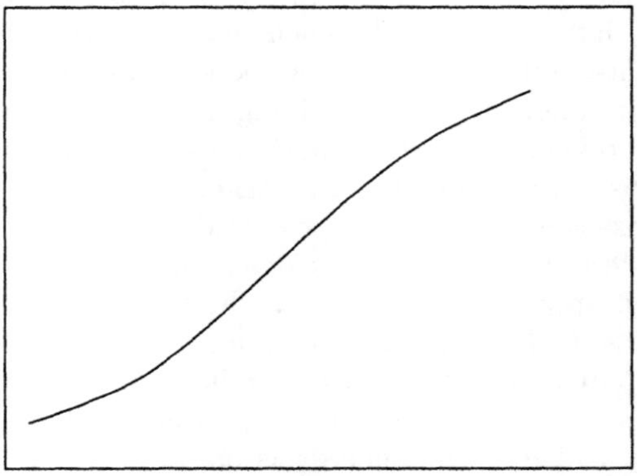

Fig. 2. Motor load vs wood-chip feeding speed [16].

10.2 Simulation Study

To simulate the proposed algorithm, it is assumed that the dynamic model of the refining system is represented by equations (1) and (2) whose parameter matrices are:

$$A = \begin{bmatrix} 0.8 \; 0.5 \\ 0.0 \; 0.9 \end{bmatrix}, \; B = \begin{bmatrix} 0.1 \\ 0.5 \end{bmatrix}, \; c = [1 \;\; 1], \; D = 1,$$
$$E = \begin{bmatrix} 0.5 \\ 0.6 \end{bmatrix}, \; F = 1, \; f_H = 10.$$

In this example, the fault is regarded as the unexpected changes of the screw transfer gain which can be directly expressed as f in equation (1). The simulation is carried out in a closed loop framework, where a state feedback control of

$$u_k = r - Kx_k \tag{116}$$

is used. The observer matrix L and the feedback matrix K are designed to be

$$L = \begin{bmatrix} 0.533 \\ 0.467 \end{bmatrix} \text{ and } K = [0.175 \;\; 0.245]$$

so that the observer eigenvalues are placed at 0.5 and 0.2, and the desired closed-loop pole locations are paced at 0.1,0.2, which are desired locations between 0 and 1. The set point is set to $r = 30$ and the two random signals, ω_k and ν, are independent truncated Gaussian processes with zero means. In the simulation, the healthy actuator has a gain $f_H = 10$ and a constant step fault is $f = 0.8f_H$ which is created at k = 50 after an initial healthy operation of the closed loop system. This fault is switched back to $f = f_H$ at sample time k = 100. A slow-drifting fault is created at sample time k = 200 with

$$f_k = \begin{cases} f_{k-1} + 0.005 f_H, \ 200 \le k < 250 \\ f_{k-1} - 0.005 f_H, \ 250 \le k < 300 \end{cases} \tag{117}$$

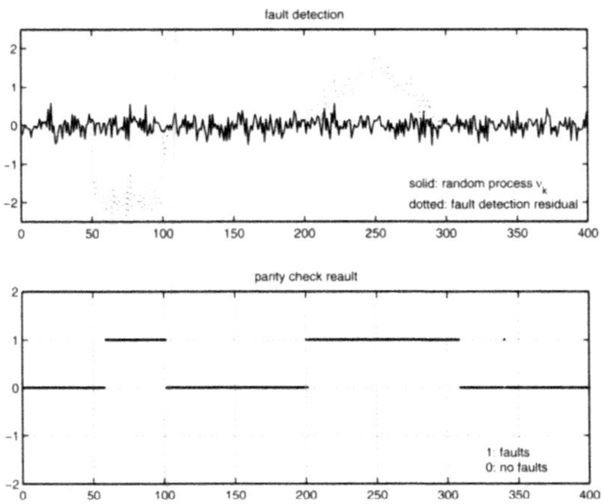

Fig. 3. Result of fault detection and diagnosis

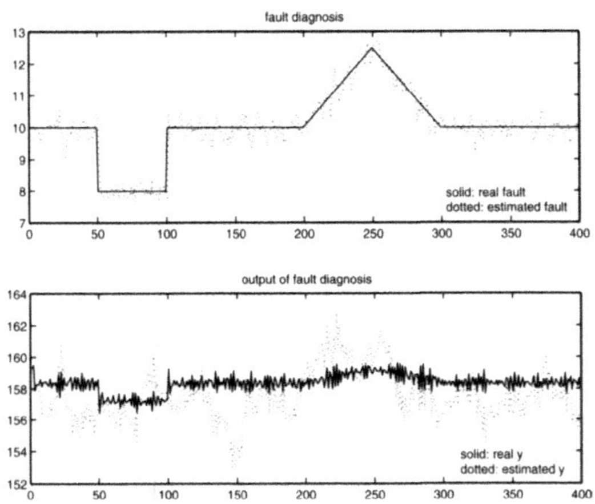

Fig. 4. State estimations of fault diagnosis

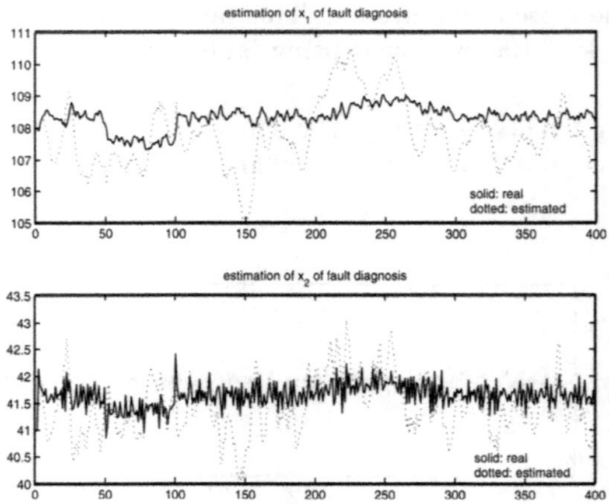

Fig. 5. Output estimation and control input

The simulation results are shown in Figs. 3–5, where in Fig. 3 the fault detection and the parity check results using (21)–(23) are given. The response of the parity check has shown a clear, but small detection delay that is in fact caused by the window length applied. The window length for the parity check algorithm is set to be $N = 10$ and the threshold is set to be $\lambda = 8$. It can be seen that the fault has been correctly detected. Fig. 4 shows the results of fault diagnosis and the response of the system and the observer outputs. In Fig. 5, the response of the system state vector and its estimation are displayed. From these figures it can be seen that the fault cannot be simply diagnosed by visual inspection of the system state/output variables and their direct estimations. This has justified the necessity and the effectiveness of using the established recursive fault diagnosis algorithm in equation (38).

11 Discussions and Conclusions

In this chapter, a new fault diagnosis algorithm has been firstly established for linear stochastic systems subjected to actuator faults. The observer based approach is used to construct the residual signals and the fault detection is performed using the statistically based parity check method. In terms of fault diagnosis, a recursive algorithm is developed. This algorithm is used to control the residual signal so that the resulting residual is only affected by the random inputs for constant faults. It has been shown that the developed fault diagnosis algorithm can also be applied to slow-drifting faults, where the resulting residual signal is related to both the random inputs and the uncontrollable part that is caused by the unknown rate of the drifting faults.

The fault diagnosis for linear systems with random parameters have also be formulated, where a new application of the Laplace transform to the probability density functions has been made. This leads to the fault diagnosis algorithm as shown in equation (92), where the scanning parameter estimation in s-domain is used. A general approach has also been developed that designs the fault diagnosis algorithm by minimizing the residual entropy for systems with a known nonlinear dynamics.

An applicability study on the mechanical pulping process is made to illustrate the use of the fault diagnosis algorithm, where a simulated example demonstrates the effectiveness of the proposed algorithms. Desired results have been obtained.

References

1. Dumont G.A. (1982) Self-Tuning Control of a Chip Refiner Motor Load. Automatica 18:307–314
2. Dumont G.A., Astrom K.J. (1988) Wood Chip Refiner Control. IEEE Control Systems Magazine 8:38–43
3. Frank P.M. (1994) On-Line Fault Detection in Uncertain Nonlinear Systems Using Diagnostic Observers: A Survey. Int. Jour. of Systems Science 25:2129–2154
4. Frank P. M. (1997) Deterministic nonlinear observer-based approaches to fault diagnosis: a survey. Control Engineering Practice 5:663 -670
5. Kabore P., Wang H. (2001) Design of Fault Diagnosis Filters and Fault Tolerant Control for a Class of Nonlinear Systems. IEEE Trans. on Automatic Control 46: 1805–1810
6. Isermann R., Balle P. (1996) Trends in the Application of Model Based Fault Detection and Diagnosis of Technical Processes. Proc. of 13th IFAC World Congress 13:1–12
7. Nikiforov I., Staroswiecki M., Vozel B. (1996) Duality of Analytical Redundancy and Statistical Approach in Fault Diagnosis. Proc. of 13th IFAC World Congress 13:19–24
8. Papoulis A. (1991) Probability, Random Variables and Stochastic Processes, Third Edition, McGraw-Hill
9. Patton R.J., Frank P., Clark R. (1989) Fault Diagnosis in Dynamic Systems: Theory and Application. Prentice-Hall, Englewood Cliffs, NJ
10. Patton R.J. (1994) Robust Model-Based Fault Diagnosis: The State of the Art. Proc. of Safe-Process'94, Espoo, Finland
11. Polycarpou M.M. (2000) Design and Stability Analysis of a Fault Accommodation Scheme for a Class of Nonlinear Dynamical Systems. Proc. of 4th IFAC Symposium on Fault Detection, Supervision and Safety for Technical Processes, Budapest, 2:733–738
12. Smook G.D. (1992) Handbook for Pulp and Paper Technologists. 1992 Angus Wilde Publications, Vancouver
13. Wang H., Daley S. (1996) Actuator Fault Diagnosis: An Adaptive Observer Based Approach. IEEE Trans. on Automatic Control 42:1073-1077

14. Wang H. (2000) Bounded Dynamic Stochastic Distributions: Modelling and Control, Springer-Verlag, London
15. Wang H. (2001) Detecting Faults in Dynamic and Bounded Stochastic Distributions: An Observer Based Technique. Proc. of 2001 American Control Conference, Virginia, USA, 482–487
16. Wang H., Wang A.P., Duncan S. (1997) Advanced Process Control in Paper and Board Making, Pira International
17. Wang Y., Wang H. (2002) Output Probability Density Function Control for Linear Stochastic Systems with Arbitrarily Bounded Random Parameters: A New Application of the Laplace Transform. Proc. of 2002 American Control Conference, Alaska, 4262–4267

Fault Diagnosis for Industrial Robots

Fabrizio Caccavale[1] and Luigi Villani[2]

[1] Dipartimento di Fisica ed Ingegneria dell'Ambiente
Università degli Studi della Basilicata
Contrada Macchia Romana, 85100, Potenza, Italy

[2] Dipartimento di Informatica e Sistemistica
Università degli Studi di Napoli Federico II
Via Claudio 21, 80125 Napoli, Italy

Abstract. In the last decade considerable research efforts have been spent to seek for systematic approaches to Fault Diagnosis (FD) in dynamical systems. Special attention has been put in robotic systems, especially for those operating in remote or hazardous environments, where a high degree of safety as well as self-diagnostics capabilities are required. On the other hand, the development of effective strategies of fault diagnosis for robot manipulators operating in an industrial context is a critical research task. Several FD techniques for robot manipulators have been proposed in the literature, although the problem of their application to industrial robots has not been extensively investigated. In this chapter different discrete-time observer-based approaches to FD for mechanical manipulators are presented and critically compared. First, a rough FD technique is considered, which is based solely on the prediction capabilities of the manipulator dynamic model. Next, an observer-based technique is presented, where a robust time-delayed compensation is introduced to cope with disturbances and modeling uncertainties. Finally, two different observer-based schemes are developed, where the uncertain terms in the model are dynamically estimated: the first scheme is based on the recursive estimation of the uncertain terms, while the second one adaptively estimates the parameters of a suitable parametric model of the uncertainties. All the considered schemes are experimentally tested on a six-degree-of-freedom industrial robot and the performance are critically compared each other.

1 Introduction

The adoption of effective fault diagnosis (FD) techniques is becoming critical to ensure higher levels of safety and reliability in automated plants and autonomous systems. In the last decade considerable research efforts have been spent to seek for systematic approaches to fault diagnosis in dynamical systems. In this framework special attention has been paid to robotic systems, especially for those operating in remote or hazardous environments, where a high degree of safety and self-diagnostics capabilities are required. However, the development and the application of FD techniques is also of the utmost importance for industrial robots, where the main objectives are the achievement of a safe man-machine interaction as well as a quick and appropriate reaction of the system to the occurrence of failures.

F. Caccavale, L. Villani (Eds.): Fault Diagnosis and Fault Tolerance for Mechatronic Systems, STAR 1, pp. 85-108, 2003.

The main goal of an FD system is the monitoring of the plant during its normal working conditions so as to detect the occurrence of failures (*fault detection*), recognize the location (*fault isolation*) and the time evolution (*fault identification*) of the failures. In the model-based approach to FD, this goal is achieved by comparing the actual system's behavior with the corresponding expected behavior derived via its mathematical model. Usually, the output of a fault detection algorithm is a set of variables sensitive to the occurrence of a failure (*residuals*). Namely, when a a failure occurs, a *fault signature* affects the residuals. Then, the information from the signatures is processed to identify the size and the location of the fault. The interested reader is referred to [10–12,4,16] for a wide overview of the existing model-based FD techniques. As for the case of nonlinear dynamical systems the fault detection methods can be roughly regrouped in three main classes: observer-based approaches [10,11], parameter estimation techniques [13,14] and algorithms based on learning methodologies [18,24,6,22,23]. Recently, soft computing methods, integrating quantitative and qualitative modeling information, have been developed to improve FD reasoning capabilities (see [17] and references therein).

Several FD techniques for mechanical manipulators have been developed in the literature, based on parameter estimation [14,9], on the combined use of state observers and fuzzy logic residuals evaluation [19] and on discrete-time observers [2,3]. Moreover, in [26] an analytical redundancy concept [5] is applied together with an adaptive update of the thresholds on the residuals, while in [25] a neural network is employed to match an unknown fault.

This chapter focuses on nonlinear observer-based FD approaches for mechanical manipulators. The goal is to develop FD schemes suitable for the application to conventional industrial manipulators, which overcome the typical problems of industrial setups, i.e., lack of knowledge of some terms in the mathematical model, effects of sensory noise and of discrete-time implementation.

Usually, the observer-based methods require a model of the system to be operated in parallel to the process (*diagnostic observer*) in an open-loop fashion [19]. Then, the residuals are computed as the difference between the measured output variables and those predicted via the diagnostic observer. Assuming an exact knowledge of the manipulator dynamics, the residuals should become nonzero when a fault occurs. However, perfect knowledge of the manipulator model is rarely a reasonable assumption, especially for industrial gear-driven robots, where effects like backlash and friction become not negligible, and yet difficult to model. Further, the diagnostic observer has to be implemented as a discrete-time system eventually working at a low sampling rate. In order to limit the computational burden, a simple discretization scheme should be used to implement the observer; this may lead to a drifting behavior of the residuals. Hence, the observer cannot be operated in an open-loop fashion: some information coming from the plant (through the sensors) must be provided to the observer in order to improve its robustness to both model uncertainties and discretization errors. However, robust detection of the faults can be achieved only at the expense of reduced sensitivity to failures; this requires that solutions based on a trade-off between these two performance indexes must be devised.

Hence, the design of a diagnostic observer must take into account two main requirements:

- In order to take into account the unavoidable effects of the discretization, the diagnostic observer should be designed directly in the discrete-time domain;
- Information provided by sensor measurements must be used, so as to improve the robustness of the observer without a destructive impact on the sensitivity to the failures.

In order to fulfill the above requirements, a diagnostic observer is designed so as to estimate the manipulator state; it includes a term depending on the state estimation error and a term compensating for unmodeled dynamics, disturbances and noise. Different strategies for computing the compensation term are considered and critically compared. Then, a suitable choice of the residuals, computed on the basis of the observer outputs, allows the design of a simple and reliable fault isolation procedure, i.e., the type and the size of the failures can be determined from the signatures on the residuals. Remarkably, the FD scheme is designed directly in discrete-time so as to take into account the effect of discretization.

Finally, the proposed FD schemes are experimentally tested on a six-degree-of-freedom industrial manipulator with open control architecture.

2 Modeling

The dynamic model of an n-degrees-of-freedom rigid robot in the continous time can be written in the form

$$\boldsymbol{M}(\boldsymbol{q}(t))\ddot{\boldsymbol{q}}(t) + \boldsymbol{n}(\boldsymbol{q}(t), \dot{\boldsymbol{q}}(t)) = \boldsymbol{\tau}(t), \tag{1}$$

where $\boldsymbol{q}$ and $\boldsymbol{\tau}$ denote the $(n \times 1)$ vectors of joint variables and joint torques, respectively. In equation (1) $\boldsymbol{M}$ is the $(n \times n)$ symmetric and positive definite inertia matrix and

$$\boldsymbol{n}(\boldsymbol{q}, \dot{\boldsymbol{q}}) = \boldsymbol{V}(\boldsymbol{q}, \dot{\boldsymbol{q}})\dot{\boldsymbol{q}} + \boldsymbol{F}\dot{\boldsymbol{q}} + \boldsymbol{g}(\boldsymbol{q}) + \boldsymbol{\nu}(\boldsymbol{q}, \dot{\boldsymbol{q}}) \tag{2}$$

is the $(n \times 1)$ vector collecting the Coriolis and centrifugal ($\boldsymbol{V}(\boldsymbol{q}, \dot{\boldsymbol{q}})\dot{\boldsymbol{q}}$), friction ($\boldsymbol{F}\dot{\boldsymbol{q}}$) and gravitational ($\boldsymbol{g}(\boldsymbol{q})$) terms, while $\boldsymbol{\nu}(\boldsymbol{q}, \dot{\boldsymbol{q}})$ collects other dynamic terms, usually difficult to model (e.g.,friction effects at low velocities, motors electromagnetic disturbances). Hereafter, it is assumed that only an approximate model of the manipulator is known, i.e., only nominal estimates

$(\widehat{\boldsymbol{M}}(\boldsymbol{q}), \widehat{\boldsymbol{n}}(\boldsymbol{q}, \dot{\boldsymbol{q}}))$ are available for the terms in (1). By introducing the quantities $\widetilde{\boldsymbol{M}} = \boldsymbol{M} - \widehat{\boldsymbol{M}}$ and $\widetilde{\boldsymbol{n}} = \boldsymbol{n} - \widehat{\boldsymbol{n}}$, the equations of motion can be rewritten in the form

$$\widehat{\boldsymbol{M}}(\boldsymbol{q}(t))\ddot{\boldsymbol{q}}(t) + \widehat{\boldsymbol{n}}(\boldsymbol{q}(t), \dot{\boldsymbol{q}}(t)) = \boldsymbol{\tau}(t) - \boldsymbol{\delta}(\boldsymbol{q}(t), \dot{\boldsymbol{q}}(t), \boldsymbol{\tau}(t)), \tag{3}$$

where the modeling uncertainties are explicitly taken into account by the term

$$\boldsymbol{\delta}(\boldsymbol{q}, \dot{\boldsymbol{q}}, \boldsymbol{\tau}) = \widetilde{\boldsymbol{M}}(\boldsymbol{q}) \left(\boldsymbol{M}(\boldsymbol{q})^{-1}(\boldsymbol{\tau} - \boldsymbol{n}(\boldsymbol{q}, \dot{\boldsymbol{q}}))\right) + \widetilde{\boldsymbol{n}}(\boldsymbol{q}, \dot{\boldsymbol{q}}). \tag{4}$$

A choice for the state variables of the system is represented by the $(2n \times 1)$ vector

$$\boldsymbol{x}(t) = \begin{bmatrix} \boldsymbol{x}_1(t) \\ \boldsymbol{x}_2(t) \end{bmatrix} = \begin{bmatrix} \boldsymbol{q}(t) \\ \dot{\boldsymbol{q}}(t) \end{bmatrix}. \tag{5}$$

The state-space equations of the manipulator are then given by

$$\begin{cases} \dot{\boldsymbol{x}}(t) = \boldsymbol{A}_c\boldsymbol{x}(t) + \boldsymbol{h}_c(\boldsymbol{x}(t)) + \boldsymbol{B}_c(\boldsymbol{x}(t))\boldsymbol{u}(t) + \boldsymbol{\eta}_c(\boldsymbol{x}(t), \boldsymbol{u}(t)) \\ \boldsymbol{y}(t) = \boldsymbol{C}\boldsymbol{x}(t), \end{cases} \tag{6}$$

where $\boldsymbol{y}$ denotes the $(p \times 1)$ output vector, $\boldsymbol{u} = \boldsymbol{\tau}$ and $\boldsymbol{C}$ is the $(p \times 2n)$ output matrix. The matrices $\boldsymbol{A}_c$ and $\boldsymbol{B}_c$ in (6) are defined as follows:

$$\boldsymbol{A}_c = \begin{bmatrix} \boldsymbol{O}_n & \boldsymbol{I}_n \\ \boldsymbol{O}_n & \boldsymbol{O}_n \end{bmatrix}, \quad \boldsymbol{B}_c(\boldsymbol{x}) = \begin{bmatrix} \boldsymbol{O}_n \\ \widehat{\boldsymbol{M}}^{-1}(\boldsymbol{x}_1) \end{bmatrix}, \tag{7}$$

where $\boldsymbol{O}_n$ denotes the $(n \times n)$ null matrix and $\boldsymbol{I}_n$ denotes the $(n \times n)$ identity matrix. If the whole state (i.e., joints positions and velocities) is measurable, then $p = 2n$ and $\boldsymbol{C} = \boldsymbol{I}_{2n}$.

The other two terms in (6), representing the nonlinear part of the dynamics, are given by

$$\boldsymbol{h}_c(\boldsymbol{x}) = \begin{bmatrix} \boldsymbol{0}_n \\ -\widehat{\boldsymbol{M}}^{-1}(\boldsymbol{x}_1)\widehat{\boldsymbol{n}}(\boldsymbol{x}_1, \boldsymbol{x}_2) \end{bmatrix}, \tag{8}$$

and

$$\boldsymbol{\eta}_c(\boldsymbol{x}, \boldsymbol{u}) = \begin{bmatrix} \boldsymbol{0}_n \\ -\widehat{\boldsymbol{M}}^{-1}(\boldsymbol{x}_1)\boldsymbol{\delta}(\boldsymbol{x}_1, \boldsymbol{x}_2, \boldsymbol{u}) \end{bmatrix}, \tag{9}$$

where $\boldsymbol{0}_n$ denotes the $(n \times 1)$ null vector.

It is assumed that the measurements of the output variables are sampled at a fixed time step T, while the input torques are assumed to be constant over each time interval $\mathcal{I}_k \equiv [kT,\ (k+1)T[$, where the integer $k \geq 0$ denotes the discrete time variable. Thus, it is worth looking for a discrete-time equivalent of the model (6). To the purpose, consider a Taylor expansion of $\boldsymbol{x}_1(t)$ and $\boldsymbol{x}_2(t)$ at $t_{k+1} = (k+1)T$, with starting point $t_k = kT$

$$\begin{cases} \boldsymbol{x}_1(k+1) = \boldsymbol{x}_1(k) + T\boldsymbol{x}_2(k) + \boldsymbol{\rho}_{d1}(k) \\ \boldsymbol{x}_2(k+1) = \boldsymbol{x}_2(k) + T\dot{\boldsymbol{x}}_2(k) + \boldsymbol{\rho}_{d2}(k), \end{cases} \tag{10}$$

where the discrete time variables t_k and t_{k+1} have been denoted simply by the integers k and $k+1$, respectively. In (10) the two quantities $\boldsymbol{\rho}_{d1}$ and $\boldsymbol{\rho}_{d2}$ represent the *local* discretization errors due to the truncation of the series; they depend linearly on the joint accelerations and jerks respectively, evaluated at suitable intermediate points in $\mathcal{I}_k$. Thus, a simple discrete-time model based on the first order Euler method, can be found

$$\begin{cases} \boldsymbol{x}(k+1) = \boldsymbol{A}\boldsymbol{x}(k) + \boldsymbol{h}(\boldsymbol{x}(k)) + \boldsymbol{B}(\boldsymbol{x}(k))\boldsymbol{u}(k) + \boldsymbol{\eta}(k, \boldsymbol{x}(k), \boldsymbol{u}(k)) \\ \boldsymbol{y}(k) = \boldsymbol{C}\boldsymbol{x}(k), \end{cases} \tag{11}$$

where

$$\boldsymbol{A} = \begin{bmatrix} \boldsymbol{I}_n & T\boldsymbol{I}_n \\ \boldsymbol{O}_n & \boldsymbol{I}_n \end{bmatrix}, \quad \boldsymbol{h} = T\,\boldsymbol{h}_c, \quad \boldsymbol{B} = T\,\boldsymbol{B}_c, \tag{12}$$

and

$$\boldsymbol{\eta} = T\,\boldsymbol{\eta}_c + \begin{bmatrix} \boldsymbol{\rho}_{d1} \\ \boldsymbol{\rho}_{d2} \end{bmatrix}. \tag{13}$$

The features of the Euler method method for numerical integration of differential equations are well-known (see, e.g., [7]): the *local* truncation errors are given by $\boldsymbol{\rho}_{d1}$ an $\boldsymbol{\rho}_{d2}$, while the *global* accumulated error is of the same order of magnitude as T. It is worth noticing that different discrete dynamic models have been proposed in the literature, ensuring better performance both for simulations and control purposes (see, e.g., the model proposed in [15]). However, these models are not so simple as the Euler-based one, and they are not tailored for *direct dynamics* purposes (i.e., for the evaluation of the manipulator state at the current step, given the state and the input at the previous step).

The classes of failures considered in this work are those of *actuator* faults and *sensor* faults. The former can be defined as the class of failures occurring either in the driving motors or, eventually, in the corresponding gear trains. This class of failures can be represented as an unknown additive disturbance on the commanded torques, hereafter referred as *nominal* torques, $\bar{\boldsymbol{u}}$. Hence, an actuator fault occurring at $t_k = kT$ results in a faulty torque input given by

$$\boldsymbol{u}(k) = \bar{\boldsymbol{u}}(k) + \boldsymbol{\delta u}(k), \tag{14}$$

where $\boldsymbol{\delta u}$ represents the unknown fault.

A sensor fault occurs when the sensor readings do not coincide with the true values of the state variables, i.e., the vector of the *nominal* measures of the state variables ($\bar{\boldsymbol{x}}$) is given by

$$\bar{\boldsymbol{x}}(k) = \boldsymbol{x}(k) + \boldsymbol{\delta x}(k), \tag{15}$$

where $\boldsymbol{\delta x}$ is the unknown fault. In the general case, i.e., both joint positions and velocities are measured, the vector sensor faults is of the form

$$\boldsymbol{\delta x}(k) = \begin{bmatrix} \boldsymbol{\delta x}_1(k) \\ \boldsymbol{\delta x}_2(k) \end{bmatrix}, \tag{16}$$

where $\boldsymbol{\delta x}_1(k)$ and $\boldsymbol{\delta x}_2(k)$ are independent.

Therefore, the nominal dynamics (11) in the presence of faults becomes

$$\bar{\boldsymbol{x}}(k+1) = \boldsymbol{A}\bar{\boldsymbol{x}}(k) + \boldsymbol{h}(\bar{\boldsymbol{x}}(k)) + \boldsymbol{B}(\bar{\boldsymbol{x}}(k))\bar{\boldsymbol{u}}(k) + \boldsymbol{\eta}(k) + \boldsymbol{f}(k), \tag{17}$$

where the dependence of $\boldsymbol{f}$ and $\boldsymbol{\eta}$ upon $\boldsymbol{x}$ and $\boldsymbol{u}$ has been dropped for notation compactness. The fault vector $\boldsymbol{f}$ represents the effect of the faults on the system dynamics, and is given by $\boldsymbol{f}(k) = \boldsymbol{f}_a(k) + \boldsymbol{f}_s(k)$, where:

$$\boldsymbol{f}_a(k) = \boldsymbol{B}(\bar{\boldsymbol{x}}(k))\boldsymbol{\delta u}(k), \tag{18}$$

and

$$\begin{aligned} \boldsymbol{f}_s(k) = {} & \boldsymbol{\delta x}(k+1) - \boldsymbol{A}\boldsymbol{\delta x}(k) + \boldsymbol{\delta h}(\bar{\boldsymbol{x}}(k), \boldsymbol{\delta x}(k)) + \\ & \boldsymbol{\delta B}(\bar{\boldsymbol{x}}(k), \boldsymbol{\delta x}(k))(\bar{\boldsymbol{u}}(k) + \boldsymbol{\delta u}(k)), \end{aligned} \tag{19}$$

where $\boldsymbol{\delta h}(\bar{\boldsymbol{x}}, \boldsymbol{\delta x}) = \boldsymbol{h}(\boldsymbol{x}) - \boldsymbol{h}(\bar{\boldsymbol{x}})$ and $\boldsymbol{\delta B}(\bar{\boldsymbol{x}}, \boldsymbol{\delta x}) = \boldsymbol{B}(\boldsymbol{x}) - \boldsymbol{B}(\bar{\boldsymbol{x}})$. It is worth noticing, that an actuator fault affects only the last n components of the fault vector $\boldsymbol{f}$, while a sensor fault influences all the $2n$ components.

In the following different fault diagnosis schemes are presented. The main goal is to design a diagnostic system able to track the behaviour of the manipulator despite the discretization errors and uncertainties in the available mathematical model. Each scheme is developed with reference to the discrete-time model (17), in which it is assumed the whole state measurable. In Section 5 the case of unmeasurable velocities will be tackled.

3 A Simple Model-Based Fault Diagnosis Scheme

A rough estimate of the fault vector $\boldsymbol{f}(k)$ can be obtained by defining the residual vector as follows:

$$\boldsymbol{r}(k+1) = \bar{\boldsymbol{x}}(k+1) - \boldsymbol{A}\bar{\boldsymbol{x}}(k) - \boldsymbol{h}(\bar{\boldsymbol{x}}(k)) - \boldsymbol{B}(\bar{\boldsymbol{x}}(k))\bar{\boldsymbol{u}}(k). \tag{20}$$

From (17) it can be simply recognized that the residual is directly affected by the uncertainties

$$\boldsymbol{r}(k+1) = \boldsymbol{\eta}(k) + \boldsymbol{f}(k). \tag{21}$$

Hence, the fault can be detected and isolated only if $\|\boldsymbol{\eta}\| \leq \eta_M$ for every k, $\boldsymbol{x}$ and $\boldsymbol{u}$ in the domain of interest. In this case, a fault occurring at the time step k_f would be detected if there exists a $k \geq k_f$ such that

$$\|\boldsymbol{f}(k)\| > \eta_M > \|\boldsymbol{\eta}(k)\|. \tag{22}$$

However, either a good estimate of η_M is not available or the uncertainties cannot be considered limited. Moreover, sensitivity to faults is heavily affected by the magnitude of η_M, i.e., the effect faults with small magnitude would be masked by uncertainties.

4 Observer-Based Fault Diagnosis

In order to overcome the drawbacks of the simple strategy proposed in the previous subsection, an observer-based scheme can be adopted. The main goal is to design a so-called *diagnostic observer*, which is able to track the behaviour of the manipulator despite the discretization errors and uncertainties in the available mathematical model.

The general structure of the observer is:

$$\widehat{\boldsymbol{x}}(k+1) = \boldsymbol{A}\widehat{\boldsymbol{x}}(k) + \boldsymbol{h}(\bar{\boldsymbol{x}}(k)) + \boldsymbol{B}(\bar{\boldsymbol{x}}(k))\bar{\boldsymbol{u}}(k) + \boldsymbol{K}_o\boldsymbol{e}(k) + \widehat{\boldsymbol{\eta}}(k), \tag{23}$$

where $\boldsymbol{e} = \bar{\boldsymbol{x}} - \widehat{\boldsymbol{x}}$ is the state estimation error, and the notation '$\widehat{\ }$' denotes the estimated variables. In equation (23) the term $\boldsymbol{K}_o\boldsymbol{e}$ performs a feedback action based on the estimation error, where

$$\boldsymbol{K}_o = \begin{bmatrix} \boldsymbol{K}_1 & T\boldsymbol{I}_n \\ \boldsymbol{O}_n & \boldsymbol{K}_2 \end{bmatrix}, \tag{24}$$

and $\boldsymbol{K}_1$ and $\boldsymbol{K}_2$ are positive definite diagonal $(n \times n)$ matrices. The term $\widehat{\boldsymbol{\eta}}(k)$ represents an estimate of the uncertainties $\boldsymbol{\eta}(k)$.

Therefore, the estimation error dynamics is given by

$$\boldsymbol{e}(k+1) = \boldsymbol{F}\boldsymbol{e}(k) + \widetilde{\boldsymbol{\eta}}(k) + \boldsymbol{f}(k), \tag{25}$$

where $\boldsymbol{F}_i = \boldsymbol{I}_n - \boldsymbol{K}_i$ $(i = 1, 2)$ and $\boldsymbol{F} = \operatorname{block\,diag}\{\boldsymbol{F}_1, \boldsymbol{F}_2\}$ and $\widetilde{\boldsymbol{\eta}}(k) = \boldsymbol{\eta}(k) - \widehat{\boldsymbol{\eta}}(k)$.

The residual vector can be chosen as:

$$\boldsymbol{r}(k+1) = \boldsymbol{e}(k+1) - \boldsymbol{F}\boldsymbol{e}(k), \tag{26}$$

which can be rewritten as

$$\boldsymbol{r}(k+1) = \widetilde{\boldsymbol{\eta}}(k) + \boldsymbol{f}(k). \tag{27}$$

It can be recognized that, differently from (20), the residual vector is affected by the fault vector and the estimation error of the uncertain term, i.e. $\widetilde{\boldsymbol{\eta}}(k)$. Hence, if an accurate estimation of $\boldsymbol{\eta}(k)$ is achieved, the fault signature on the residual becomes more obvious.

In the following, three different estimation techniques for $\boldsymbol{\eta}(k)$ are presented.

4.1 Time-Delayed Estimation

The estimation the uncertain term in equation (17) can be thought as a problem of unknown input estimation. In this framework, the so-called time-delayed feedback approach [27] can be effectively adopted; moreover, the results obtained in [21] for linear systems are closely related both to our problem and to time-delayed feedback techniques.

Therefore, an estimate of the uncertain term can be obtained by resorting to a time-delayed evaluation of the the dynamics of the system [2,3]. Namely, $\widehat{\boldsymbol{\eta}}(k)$ can be computed as

$$\widehat{\boldsymbol{\eta}}(k) = \bar{\boldsymbol{x}}(k) - \boldsymbol{A}\bar{\boldsymbol{x}}(k-1) - \boldsymbol{h}(\bar{\boldsymbol{x}}(k-1) - \boldsymbol{B}(\bar{\boldsymbol{x}}(k-1))\bar{\boldsymbol{u}}(k-1), \tag{28}$$

which can be rewritten as

$$\widehat{\boldsymbol{\eta}}(k) = \boldsymbol{\eta}(k-1) + \boldsymbol{f}(k-1). \tag{29}$$

Hence, the estimation error dynamics becomes

$$\boldsymbol{e}(k+1) = \boldsymbol{F}\boldsymbol{e}(k) + \boldsymbol{\eta}(k) - \boldsymbol{\eta}(k-1) + \boldsymbol{f}(k) - \boldsymbol{f}(k-1), \tag{30}$$

while the residual is given by

$$\boldsymbol{r}(k+1) = \boldsymbol{\eta}(k) - \boldsymbol{\eta}(k-1) + \boldsymbol{f}(k) - \boldsymbol{f}(k-1). \tag{31}$$

Differently from (21), the effect of the uncertainties on the residual does not depend on the magnitude of $\boldsymbol{\eta}(k)$, but on its rate of change. If $\boldsymbol{\eta}(k)$ can be assumed slowly varying, i.e.:

- its rate of change is upper bounded by a constant $\Delta\eta_M$:

$$\|\boldsymbol{\eta}(k) - \boldsymbol{\eta}(k-1)\| < \Delta\eta_M$$

  for every $\boldsymbol{x}$ and $\boldsymbol{u}$ in the domain of interest;
- the upper bound $\Delta\eta_M$ is lower than η_M:

$$\eta_M < \Delta\eta_M,$$

Hence, the residuals are less sensitive to uncertainties with respect to (21), i.e., a fault occurring at the time step k_f would be detected if there exists a $k \geq k_f$ such that

$$\|\boldsymbol{f}(k) - \boldsymbol{f}(k-1)\| > \Delta\eta_M > \|\boldsymbol{\eta}(k) - \boldsymbol{\eta}(k-1)\|. \tag{32}$$

In other words, the time-delayed approach is capable of detecting only abrupt faults, i.e., faults such that the rate of change of $\boldsymbol{f}$ is lower bounded by the maximum norm of the rate of change of $\boldsymbol{\eta}$.

4.2 Recursive Estimation

As already pointed out, the time-delayed estimation strategy does not perform well in the presence of incipient faults, i.e., such that $\boldsymbol{f}$ is characterized by a lower rate of change with respect $\boldsymbol{\eta}$. To overcome this drawback, the estimation of the uncertain term can be performed in a recursive way. Namely, $\widehat{\boldsymbol{\eta}}(k)$ can be computed as

$$\widehat{\boldsymbol{\eta}}(k+1) = \widehat{\boldsymbol{\eta}}(k) + \boldsymbol{\Gamma}_\eta\left(\boldsymbol{e}(k+1) - \boldsymbol{F}\boldsymbol{e}(k)\right), \tag{33}$$

where $\boldsymbol{\Gamma}_\eta$ is chosen such that $\boldsymbol{I}_p - \boldsymbol{\Gamma}_\eta$ has all its eigenvalues inside the unit circle.

In this case, the estimation error dynamics is given by

$$\begin{cases} \boldsymbol{e}(k+1) = \boldsymbol{F}\boldsymbol{e}(k) + \widetilde{\boldsymbol{\eta}}(k) + \boldsymbol{f}(k) \\ \widetilde{\boldsymbol{\eta}}(k+1) = (\boldsymbol{I}_p - \boldsymbol{\Gamma}_\eta)\widetilde{\boldsymbol{\eta}}(k) + \boldsymbol{\eta}(k+1) - \boldsymbol{\eta}(k) - \boldsymbol{\Gamma}_\eta \boldsymbol{f}(k). \end{cases} \tag{34}$$

Therefore, by taking into account (27), the residuals vector can be expressed as

$$\boldsymbol{r}(k+1) = (\boldsymbol{I}_p - \boldsymbol{\Gamma}_\eta)\boldsymbol{r}(k) + \boldsymbol{\eta}(k) - \boldsymbol{\eta}(k-1) + \boldsymbol{f}(k) - \boldsymbol{f}(k-1). \tag{35}$$

It can be easily recognized that the residual in (35) is a low-pass filtered version of the residual generated via (31). Hence, by suitably choosing the matrix $\boldsymbol{\Gamma}_\eta$, the scheme may be capable of detecting incipient faults.

4.3 Adaptive Estimation

The approaches described above do not make use of any *a priori* information on the uncertainties. As a matter of fact, a parametric model of the uncertainties is often available, although some parameters in the model might be unknown. In this case an adaptive estimation algorithm of the unknown parameters can be set up. It is worth remarking that such a paradigm has been keenly exploited for adaptive fault identification (see, e.g., the work in [18]–[23]). However, in this work the same concept is exploited in order to adaptively compensate for the uncertainties, so as to obtain small values of the residuals in the absence of faults.

It is assumed that the uncertain term is a function of state, input and p constant parameters

$$\boldsymbol{\eta}(k) = \boldsymbol{\eta}(k, \boldsymbol{x}(k), \boldsymbol{u}(k), \boldsymbol{\theta}), \tag{36}$$

where $\boldsymbol{\theta}$ is the $(p \times 1)$ parameters vector.

In this case, the uncertain term can be indirectly evaluated through the estimation of $\boldsymbol{\theta}$. Namely, and adaptive update law for the parameters estimate $\boldsymbol{\theta}(k)$ can be chosen as

$$\widehat{\boldsymbol{\theta}}(k+1) = \widehat{\boldsymbol{\theta}}(k) + \boldsymbol{Z}^T(k, \widehat{\boldsymbol{\theta}}(k))\boldsymbol{\Gamma}_\theta(k)\left(\boldsymbol{e}(k+1) - \boldsymbol{F}\boldsymbol{e}(k)\right), \tag{37}$$

where the $(2n \times p)$ Jacobian matrix $\boldsymbol{Z}$ is given by

$$\boldsymbol{Z}(k, \boldsymbol{\theta}) = \frac{\partial \boldsymbol{\eta}(k, \boldsymbol{x}(k), \boldsymbol{u}(k), \boldsymbol{\theta})}{\partial \boldsymbol{\theta}}. \tag{38}$$

The gain matrix $\boldsymbol{\Gamma}_\theta(k)$ is chosen as follows:

$$\boldsymbol{\Gamma}_\theta(k) = 2\left(\boldsymbol{Z}(k, \widehat{\boldsymbol{\theta}}(k))\boldsymbol{Z}^T(k, \widehat{\boldsymbol{\theta}}(k)) + \boldsymbol{Q}\right)^{-1}, \tag{39}$$

where $\boldsymbol{Q}$ is a positive definite symmetric matrix.

By resorting to a Taylor expansion about $(\boldsymbol{x}(k), \boldsymbol{u}(k), \widehat{\boldsymbol{\theta}}(k))$, the estimation error can be expressed as

$$\begin{aligned} \widetilde{\boldsymbol{\eta}}(k) &= \boldsymbol{\eta}(k, \boldsymbol{x}(k), \boldsymbol{u}(k), \boldsymbol{\theta}) - \boldsymbol{\eta}(k, \boldsymbol{x}(k), \boldsymbol{u}(k), \widehat{\boldsymbol{\theta}}(k)) \\ &\simeq \boldsymbol{Z}(k, \widehat{\boldsymbol{\theta}}(k))\widetilde{\boldsymbol{\theta}}(k) + \boldsymbol{\epsilon}(k), \end{aligned} \tag{40}$$

where $\boldsymbol{\epsilon}(k)$ is the error due to the truncation of the series and $\widetilde{\boldsymbol{\theta}}(k) = \boldsymbol{\theta} - \widehat{\boldsymbol{\theta}}(k)$ is the parameters estimation error.

The estimation error dynamics is then given by

$$\begin{cases} \boldsymbol{e}(k+1) = \boldsymbol{F}\boldsymbol{e}(k) + \boldsymbol{Z}(k, \widehat{\boldsymbol{\theta}}(k))\widetilde{\boldsymbol{\theta}}(k) + \boldsymbol{f}(k) + \boldsymbol{\epsilon}(k) \\ \widetilde{\boldsymbol{\theta}}(k+1) = \left(\boldsymbol{I}_p - \boldsymbol{Z}^T(k, \widehat{\boldsymbol{\theta}}(k))\boldsymbol{\Gamma}_\theta(k)\boldsymbol{Z}(k, \widehat{\boldsymbol{\theta}}(k))\right)\widetilde{\boldsymbol{\theta}}(k) \\ \qquad\qquad -\boldsymbol{Z}^T(k, \widehat{\boldsymbol{\theta}}(k))\boldsymbol{\Gamma}_\theta(k)\left(\boldsymbol{f}(k) + \boldsymbol{\epsilon}(k)\right). \end{cases} \tag{41}$$

Therefore, the residual becomes

$$\boldsymbol{r}(k+1) = \boldsymbol{Z}(k, \widehat{\boldsymbol{\theta}}(k))\widetilde{\boldsymbol{\theta}}(k) + \boldsymbol{f}(k) + \boldsymbol{\epsilon}(k). \tag{42}$$

It can be recognized that residuals vector is influenced by the parameters estimation error and the error due to the truncation of the series. Moreover, the influence of the past values of $\boldsymbol{f}(k)$ on $\boldsymbol{r}(k+1)$ is filtered out by the dynamics of the parameters estimation error.

5 Observer-Based Fault Diagnosis in the Absence of Velocity Measurements

If velocity measurements are not available the joint velocities are usually obtained via numerical reconstruction, e.g., using a first order difference of the measured joint positions

$$\boldsymbol{z}_2(k) = \frac{\boldsymbol{x}_1(k) - \boldsymbol{x}_1(k-1)}{T}, \tag{43}$$

where the relationship with the true time derivative can be expressed as $\boldsymbol{x}_2(k) = \boldsymbol{z}_2(k) + \boldsymbol{\rho}_{2z}(k-1)$, being $\boldsymbol{\rho}_{2z}(k-1)$ an discretization error term depending on the chosen discretization method (e.g., the first-order backward difference as in (43)).

In this case the model (11) can be rewritten in terms of the numerically reconstructed velocities rather than the actual velocities, and a new discrete-time model can be derived, which is formally similar to (11)

$$\begin{cases} \boldsymbol{z}(k+1) = \boldsymbol{A}\boldsymbol{z}(k) + \boldsymbol{h}(\boldsymbol{z}(k)) + \boldsymbol{B}(\boldsymbol{z}(k))\boldsymbol{u}(k) + \boldsymbol{\eta}_z(k, \boldsymbol{z}(k), \boldsymbol{u}(k)) \\ \boldsymbol{y}_z(k) = \boldsymbol{z}(k), \end{cases} \tag{44}$$

where

$$\boldsymbol{z}(k) = \begin{bmatrix} \boldsymbol{z}_1(k) \\ \boldsymbol{z}_2(k) \end{bmatrix}, \qquad \boldsymbol{z}_1(k) = \boldsymbol{x}_1(k), \tag{45}$$

and $\boldsymbol{\eta}_z$ is a suitably defined vector depending on $\boldsymbol{\eta}$ and $\boldsymbol{\rho}_{2z}$. Therefore, the approach described in the following for the design of the diagnostic observer can be extended to this case. Clearly, the new model is written in terms of a new set of variables, in which the actual joint velocities are replaced with their numerical counterparts; this is not to be considered as a drawback, since the main goal of an FD scheme is not the state estimation.

Therefore, when only joint position sensors are available, the model (44) have to be considered. In this case the faulty state vector is

$$\bar{\boldsymbol{z}}(k) = \boldsymbol{z}(k) + \boldsymbol{\delta z}(k), \tag{46}$$

where

$$\boldsymbol{\delta z}(k) = \begin{bmatrix} \boldsymbol{\delta z}_1(k) \\ \boldsymbol{\delta z}_2(k) \end{bmatrix}, \tag{47}$$

where $\boldsymbol{\delta z}_1(k)$ is the vector of the position sensors faults, and

$$\boldsymbol{\delta z}_2(k) = \frac{\boldsymbol{\delta z}_1(k) - \boldsymbol{\delta z}_1(k-1)}{T}. \tag{48}$$

Hence, differently from (16), the last n components of $\boldsymbol{z}$ (i.e., $\boldsymbol{z}_2$) are not independent from the first n (i.e., $\boldsymbol{z}_1$).

6 Fault Detection, Isolation, and Identification

Once the residuals vector $\boldsymbol{r}(k)$ is computed at each step, a fault is declared if each component of $\boldsymbol{r}(k)$ exceeds a suitably selected threshold ρ_i:

$$|r_i(k)| > \rho_i \qquad i = 1, \dots, n. \tag{49}$$

The *a priori* selection of each threshold should be based on the expressions of the residuals (21),(31),(35) and (42). Namely, proper setting of the thresholds requires an accurate knowledge of the uncertainties influence on the residuals, i.e,. of the constant valuse η_M and $\Delta\eta_M$. However, this approach often leads to extremely conservative results.

Therefore, an empirical approach may be pursued to set the residuals thresholds in alternative to (or in combination with) the approach based on the *a priori* knowledge of the uncertainties. Namely, a number of experiments in the absence of faults may be performed and the corresponding residuals recorded; then, the thresholds can be set on the basis of the maximum absolute values of each component of the residuals vector. Of course, the experimental trials should be chosen following the worst-case criterium for the residuals, i.e., the uncertainties influence should be the maximum possible. In the case of a robot manipulator, the experimental data should be collected along trajectories at very low velocities (so as to emphasize effects like friction and motor electromagnetic disturbances) as well as trajectories characterized by high acceleration and wide joint displacements (so as to emphasize the effect of inertial and configuration-dependent terms).

On the other hand, a complete fault diagnosis scheme should ensure not only the early and reliable detection of the failures, but also the isolation of the fault, i.e., the location of the failure. The expressions of the residuals vector (21),(31),(35) and (42) clearly show that the signatures of the faults reflect the structure of the fault vector $\boldsymbol{f}(k)$. In other words, it can be stated that:

- a fault on the ith actuator affects only the last n components of $\boldsymbol{r}(k)$;
- a fault on the ith sensor affects the ith component of $\boldsymbol{r}(k)$ and, possibly, the last n components (see equation (47)).

Hence, different faults correspond to distinct fault signature on the residuals: this implies that a reliable fault isolation is always ensured.

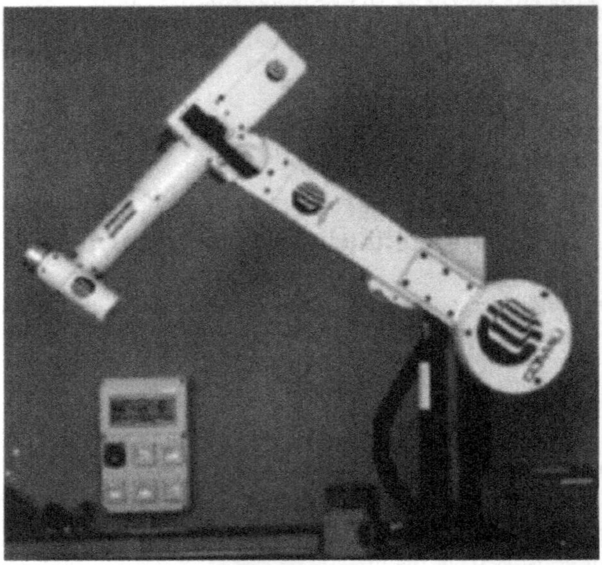

Fig. 1. The robot Comau Smart-3 S.

The problem of fault identification (i.e., the determination of the fault time evolution as accurately as possible) is a difficult task, since only the combined effect of uncertainties and faults can be estimated and not the two contribution separately. In other words, the uncertainties and the faults affect the estimation error dynamics in the same way, thus making impossible a clear distinction between faults and uncertainties influence.

In fact, the observer-based approach described above is based on the estimation of the uncertain term $\boldsymbol{\eta}(k)$. However, in the presence of faults the

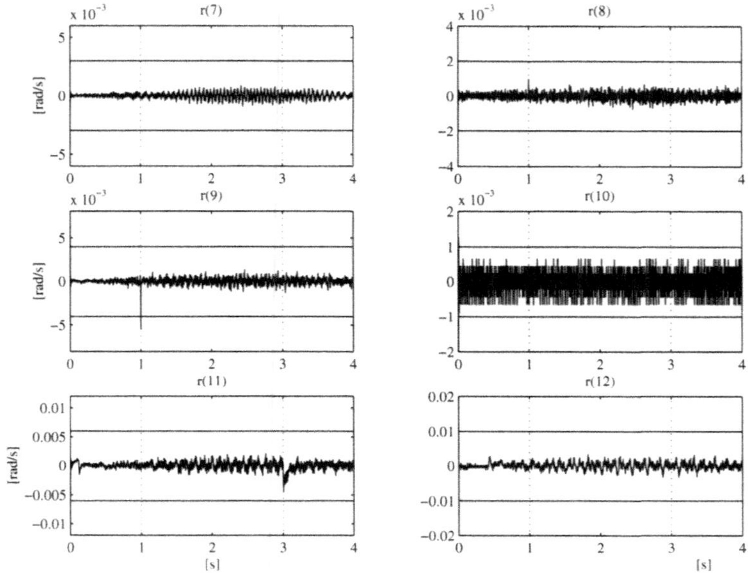

Fig. 2. Actuator fault. Time-delayed uncertainties estimation: Residuals.

estimated uncertainties $\widehat{\boldsymbol{\eta}}(k)$ will be affected a combination of $\boldsymbol{\eta}(k)$ and $\boldsymbol{f}(k)$. Therefore, the best estimate of the fault vector can be obtained only by taking $\widehat{\boldsymbol{f}}(k) = \widehat{\boldsymbol{\eta}}(k)$. In detail:

- before a fault is declared (i.e., all the components of $\boldsymbol{r}(k)$ are below the chosen thresholds), $\widehat{\boldsymbol{f}}(k)$ is set to the null vector;
- after the detection of a fault (i.e., some components of $\boldsymbol{r}(k)$ exceed the corresponding thresholds), the corresponding components of the fault vector are set equal to those of $\widehat{\boldsymbol{\eta}}(k)$.

Then, after the detection the fault can be determined as follows:

- the estimated ith component of $\boldsymbol{\delta\tau}(k)$ is set equal to the ith component of the vector $(\boldsymbol{M}(\boldsymbol{q}(k))/T)\boldsymbol{f}_2(k)$;
- the estimated ith component of $\boldsymbol{\delta x}_1(k)$ is set equal to the ith component of $\boldsymbol{x}_{1d}(k) - \bar{\boldsymbol{x}}_1(k)$,

where $\boldsymbol{f}_2$ is the vector collecting the last n components of the fault vector, $\boldsymbol{x}_{1d}$ is the commanded (desired) joint position trajectory and $\boldsymbol{e}_1$ is the vector collecting the first n components of the estimation error vector.

7 Experimental Results

The FD techniques described above are tested on the setup available in the laboratory. The robot is a Comau SMART-3 S industrial unit (Fig. 1). The

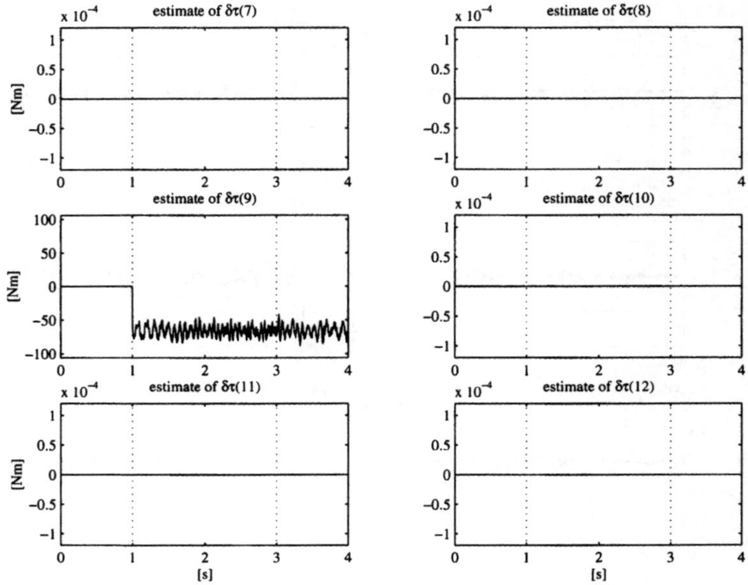

Fig. 3. Actuator fault. Time-delayed uncertainties estimation: Estimate of the fault time evolution.

manipulator has a six-revolute-joint anthropomorphic geometry with nonnull shoulder and elbow offsets and non-spherical wrist. The joints are actuated by brushless motors via gear trains; shaft absolute resolvers provide motor position measurements. The robot is controlled by the C3G 9000 control unit which has a VME-based architecture with 2 processing boards (Robot CPU and Servo CPU) both based on a Motorola 68020/68882. It is worth remarking that the SMART-3 S is a conventional industrial unit and not a research prototype; hence, all the typical drawbacks of industrial manipulators (e.g., joint friction, stiction and backlash due to the gear trains, disturbances on the torque delivered by the actuators, unmodeled elasticity of the joint shafts).

An open version of the control unit has been developed which allows testing of advanced control algorithms on a conventional industrial robot. Communication between the VME bus of the C3G 9000 unit and the ISA bus of a standard PC is made possible by a bus-to-bus adapter board and a shared memory area available in the Robot CPU. Time synchronization is implemented by interrupt signals from the C3G to the PC with data exchange at a given sampling rate. A set of C routines are available to drive the bus adapter boards. At present, a PC Pentium MMX/233 is used as control unit.

Various operational modes are available in the control unit, allowing the PC to interact with the original controller both at trajectory generation level and at joint control level. To implement model-based control schemes, the

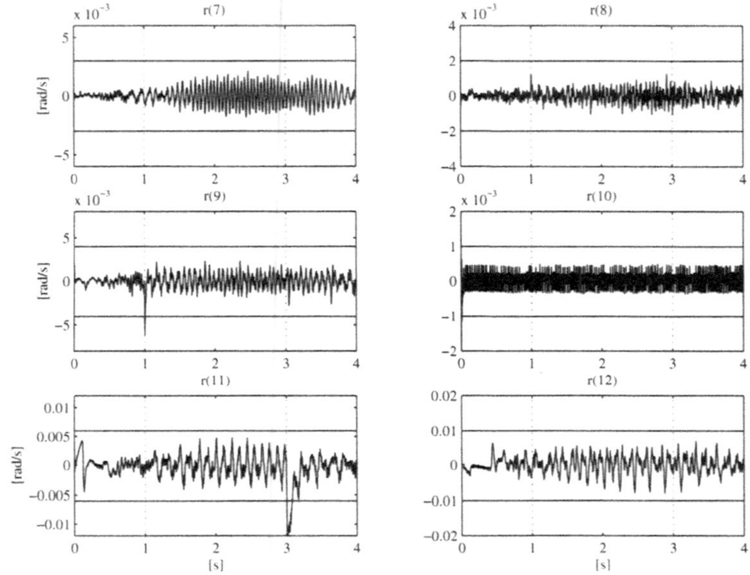

Fig. 4. Actuator fault. Recursive uncertainties estimation: Residuals.

operational mode 4 is used in which the PC is in charge of computing the control algorithm and passing the references to the current servos through the communication link.

The dynamic model of the manipulator is expressed in terms of a minimum set of dynamic parameters, estimated on the basis of direct measurements on the manipulator [1]. In the experiments a joint-space inverse dynamics control algorithm [20] is adopted working at a sampling rate of 1000 Hz.

In the experiments a fifth-order polynomial trajectory is imposed at each joint with null initial and final velocities and accelerations; the initial joint configuration is

$$\boldsymbol{q}_0^T = [\,\pi/2 \quad -2\pi/3 \quad \pi/6 \quad -\pi/2 \quad \pi/2 \quad 0.0\,]\,,$$

and the commanded joint displacement is given by:

$$\Delta\boldsymbol{q}^T = [\,\pi/4 \quad \pi/4 \quad -\pi/4 \quad 0.0 \quad -\pi/2 \quad \pi/2\,]\,.$$

The total (programmed) duration of the motion is 4 s.

The trajectory has been then executed. In order safely to emulate the presence of sensor and actuator faults, an additive signal has been superimposed to the measured experimental data off-line. Namely, the sequences $\boldsymbol{\delta\tau}(k)$ and $\boldsymbol{\delta x}_1(k)$ have been simply added to the measured fault-free data $\boldsymbol{\tau}(k)$ and $\boldsymbol{x}_1(k)$. In detail, two actuator faults have been considered affecting

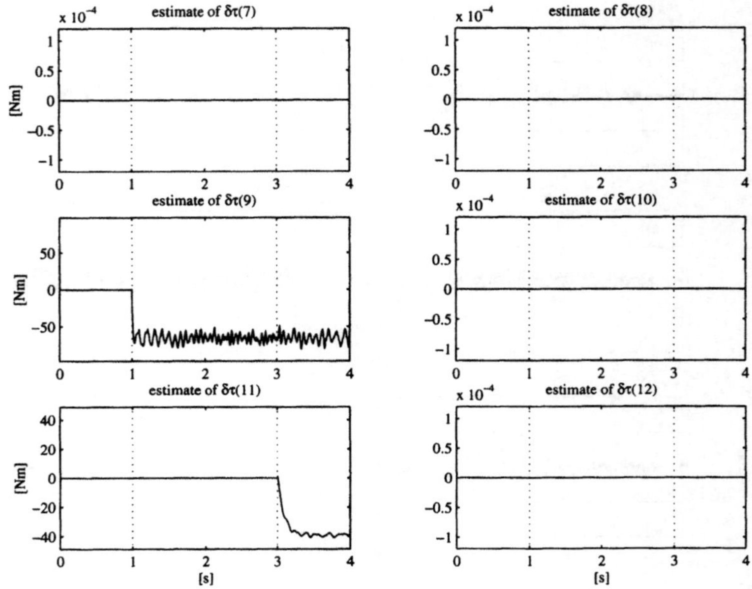

Fig. 5. Actuator fault. Time-delayed uncertainties estimation: Estimate of the fault time evolution.

the driving torques generated by the actuators of the joints 3 (occurring at time $t_{fault} = 1$ s) and 5 (occurring at time $t_{fault} = 3$ s), with the following time profile:

$$\begin{cases} \delta\tau_3(k) = 60\,(1 - \mathrm{e}^{-(kT-1)/0.002}) & kT \geq 1 \\ \delta\tau_5(k) = 40\,(1 - \mathrm{e}^{-(kT-3)/0.08}) & kT \geq 3. \end{cases}$$

Moreover, a second case study has been considered in which two sensor faults have been added to the measured angular positions of joints 3 (occurring at time $t_{fault} = 1$ s) and 5 (occurring at time $t_{fault} = 3$ s), with the following time profile:

$$\begin{cases} \delta x_{1,3}(k) = 0.006\,(1 - \mathrm{e}^{-(kT-1)/0.002}) & kT \geq 1 \\ \delta x_{1,5}(k) = 0.004\,(1 - \mathrm{e}^{-(kT-3)/0.08}) & kT \geq 3. \end{cases}$$

The first fault has to be considered as a an abrupt fault, while the second can be seen as an incipient fault.

The FD schemes are implemented at a sampling rate of 500 Hz ($T = 2$ ms). The matrix gains of the diagnostic observer (23) have been chosen as

$$\boldsymbol{K}_1 = \boldsymbol{K}_2 = 0.1\,\boldsymbol{I}_3,$$

The matrix gains in (33) and (39)are chosen as

$$\boldsymbol{\Gamma}_\eta = 0.1\mathrm{diag}\{0.1, 0.1, 0.1, 0.1, 0.1, 0.1, 2, 2, 2, 2, 2, 3\}, \qquad \boldsymbol{Q} = 2\,\boldsymbol{\Gamma}_\eta^{-1}.$$

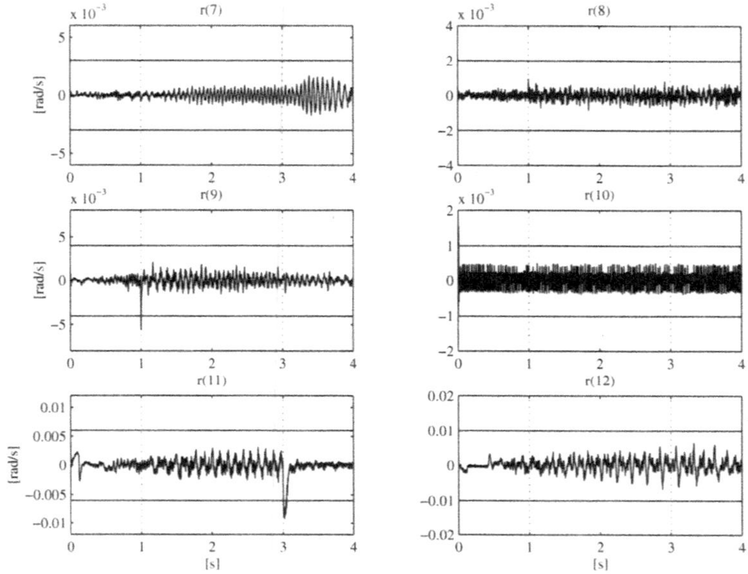

Fig. 6. Actuator fault. Adaptive uncertainties estimation: Residuals.

In order to perform a proper fault detection, suitably defined thresholds on the residuals has been selected. Thresholds setting has been achieved by measuring the residuals obtained in a set of fault-free trajectories under various operating conditions. By inspecting the measured residuals and their deviations from zero, the threshold in (49) have been set as follows:

$$\begin{aligned} &\rho_1 = 3 \cdot 10^{-5}, \quad \rho_2 = 2 \cdot 10^{-5}, \rho_3 = 4 \cdot 10^{-5}, \quad \rho_4 = 2 \cdot 10^{-5}, \\ &\rho_5 = 10^{-4}, \qquad \rho_6 = 2 \cdot 10^{-4}, \rho_7 = 3 \cdot 10^{-3}, \quad \rho_8 = 2 \cdot 10^{-3}, \\ &\rho_9 = 2.5 \cdot 10^{-3}, \rho_{10} = 10^{-3}, \quad \rho_{11} = 5 \cdot 10^{-3}, \rho_{12} = 10^{-2}. \end{aligned}$$

As for the choice of the parametric model of the uncertainties (36), it is possible to resort to different approaches. A widely adopted choice is represented by the so-called on-line approximators [18]–[22], e.g., RBF neural networks and polynomials. However, in the case of industrial robots, the uncertainties model is usually known (e.g., friction at low velocities, periodic torque disturbances), but the corresponding parameters are difficult to identify. In fact, for the robot used in the experiments an accurate dynamic model can be obtained [1] except for the periodic torque disturbances due to elec-

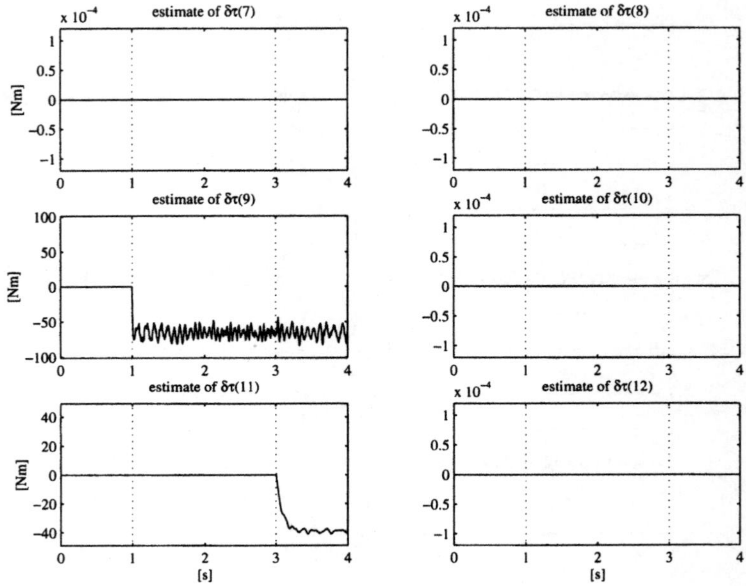

Fig. 7. Actuator fault. Adaptive uncertainties estimation: Estimate of the fault time evolution.

tromagnetic phenomena in the brushless motors. Hence, a realistic model of the uncertainties is given by [8]:

$$\begin{cases} \eta_i(k,\boldsymbol{\theta}) &= \theta_i, \\ \eta_{i+6}(k,\boldsymbol{\theta}) &= \theta_{7i} + \theta_{7i+1} sin(\theta_{7i+2} x_{1i}(k) + \theta_{7i+3} \\ & \quad + \tau_i(k) \theta_{7i+4} * sin(\theta_{7i+5} x_{i1}(k) + \theta_{7i+6}). \end{cases} \qquad i = 1, \ldots, 6 \qquad (50)$$

Suitable initial values for all parameters have been found via spectral analysis of the residuals obtained in a set of fault-free trajectories.

In the following the results obtained by adopting the observer-based approach to FD, with the different uncertainties estimation techniques, will be briefly commented.

7.1 Diagnosis Results for Sensor Faults

Figures from 2 to 7 show the obtained results in the presence of the emulated actuator fault. As expected, the time-delayed estimation technique presented in Section 4.1 is characterized by a lower sensitivity to incipient faults in front of very low residuals (see Figure 2). On the other hand, the FD scheme based on the recursive estimation of the uncertainties presented in 4.1, achieves higher sensitivity to the occurrence of incipient faults, although the residuals

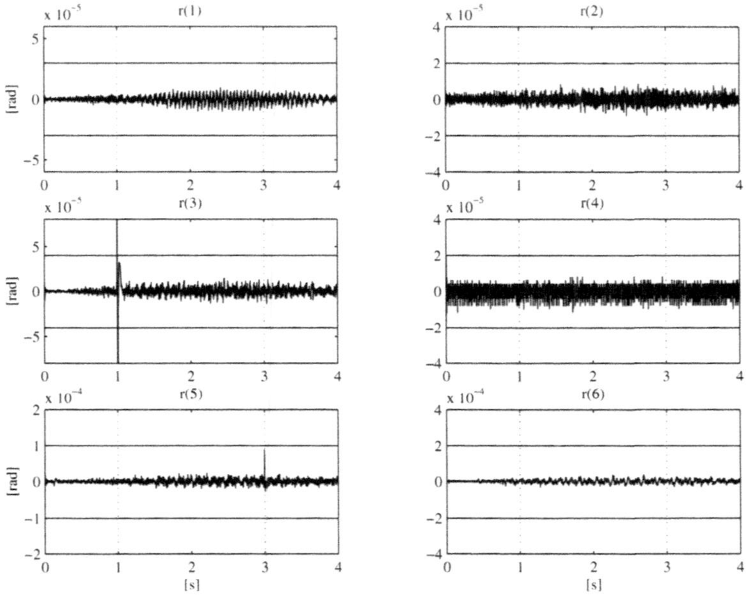

Fig. 8. Sensor fault. Time-delayed uncertainties estimation: Residuals.

magnitude is much higher (see Figure 4). The best trade-off between robustness to uncertainties (i.e., low residuals) and sensitivity to faults is achieved by adopting the adaptive estimation technique devised in Section 4.3 (see Figure 6); this result is due to the accurate choice of the uncertainties model.

Once a fault is detected, it can be reliably isolated, since the sole residuals corresponding to the faulty actuators become larger than the corresponding thresholds .

Finally, it is worth noticing that a fairly accurate fault identification can be achieved by adopting the strategies proposed in Section 6 (see Figures 3, 5 and 7).

7.2 Diagnosis Results for Actuator Faults

Figures from 8 to 13 show the obtained results in the presence of the emulated actuator fault. Again, the time-delayed estimation technique presented in Section 4.1 is characterized by a lower sensitivity to incipient faults in front of very low residuals (see Figure 8). On the other hand, the FD scheme based on the recursive estimation of the uncertainties presented in 4.1, achieves higher sensitivity to the occurrence of incipient faults, although the residuals magnitude is much higher (see Figure 10). The best trade-off between robustness to uncertainties (i.e., low residuals) and sensitivity to faults is achieved

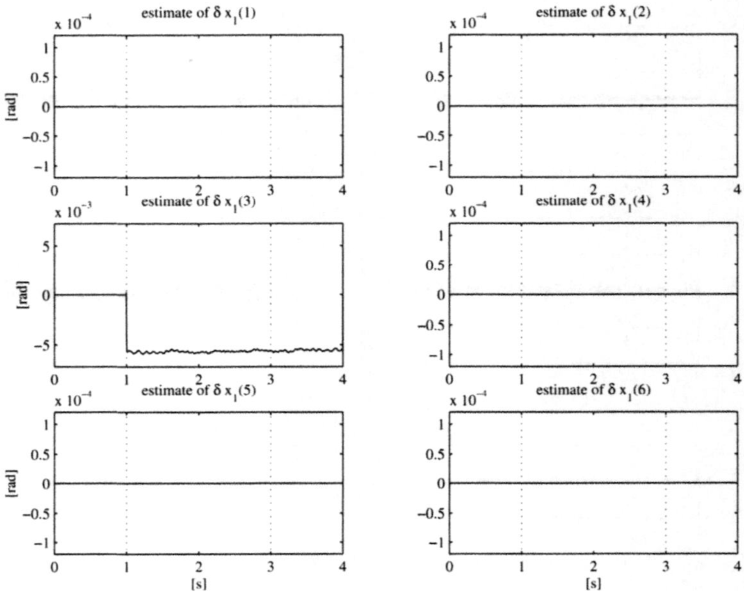

Fig. 9. Sensor fault.Time-delayed uncertainties estimation: Estimate of the fault time evolution.

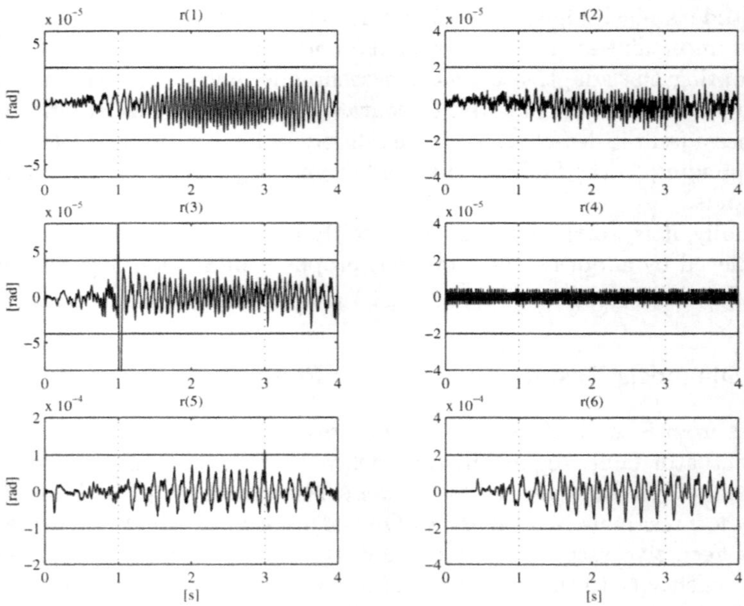

Fig. 10. Sensor fault. Recursive uncertainties estimation: Residuals.

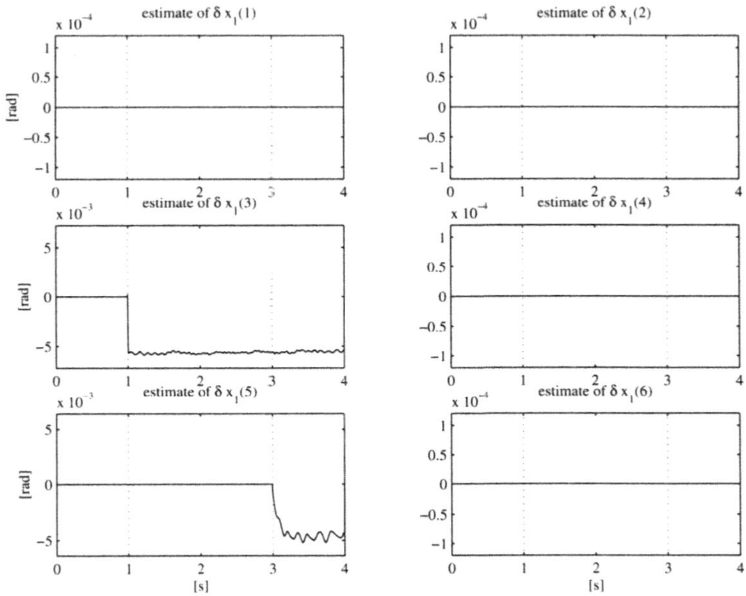

Fig. 11. Sensor fault. Time-delayed uncertainties estimation: Estimate of the fault time evolution.

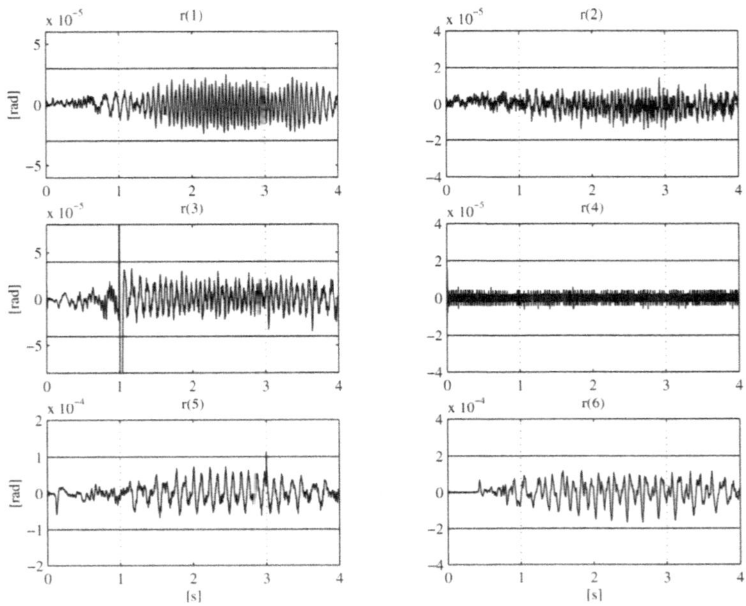

Fig. 12. Sensor fault. Adaptive uncertainties estimation: Residuals.

by adopting the adaptive estimation technique devised in Section 4.3 (see Figure 12); this result is due to the accurate choice of the uncertainties model.

Once a fault is detected, it can be reliably isolated, since the sole residuals corresponding to the faulty sensors become larger than the corresponding thresholds .

Finally, it is worth noticing that a fairly accurate fault identification can be achieved by adopting the strategies proposed in Section 6 (see Figures 9, 11 and 13).

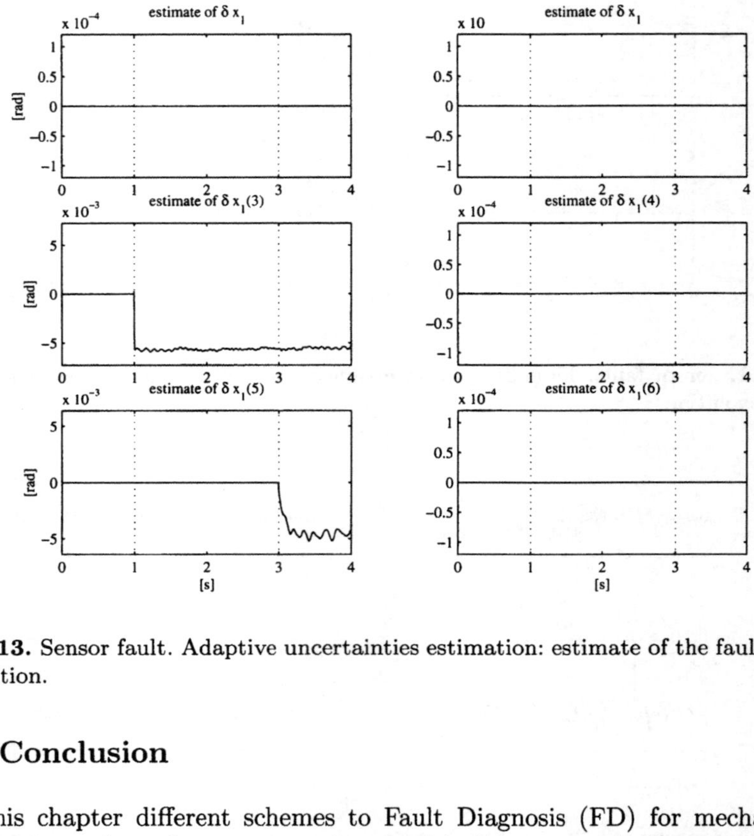

Fig. 13. Sensor fault. Adaptive uncertainties estimation: estimate of the fault time evolution.

8 Conclusion

In this chapter different schemes to Fault Diagnosis (FD) for mechanical manipulators have been presented and critically compared. Namely, all the proposed FD schemes are based on a suitably designed discrete-time observer of the manipulator's state, while each scheme differs from others for the uncertainties estimation strategy. First, a robust time-delayed estimation technique has been presented. Then, a recursive estimation of the uncertain terms is proposed which overcomes some limitations of the first technique. Finally, an adaptive estimation technique has been investigated, which makes use of a parametric model of the uncertainties. All the considered schemes have been experimentally tested on a six-degree-of-freedom industrial robot and the performance have been critically compared each other.

References

1. Antonelli G., Caccavale F., Chiacchio P. (1999) A Systematic Procedure for the Identification of Dynamic Parameters of Robot Manipulators. Robotica 17:427–435
2. Caccavale F., Walker I.D. (1997) Observer-Based Fault Detection for Robot Manipulators. Proc. of 1997 IEEE International Conference on Robotics and Automation, Albuquerque, NM, 2881–2887
3. Caccavale F. (1998) Experiments of Observer-based Fault Detection for an Industrial Robot. Proc. of 1998 IEEE International Conference on Control Applications, Trieste, I, 480–484
4. Chen J., Patton R.J. (1999) Robust Model Based Fault Diagnosis for Dynamic Systems. Kluwer Academic Publishers, Boston
5. Chow E.Y., Willsky A.S. (1984) Analytical Redundancy and the Design of Robust Failure Detection Systems. IEEE Transactions on Automatic Control 29:603–614
6. Demetriou M.A., Polycarpou M.M. (1998) Incipient Fault Diagnosis of Dynamical Systems Using Online Approximators. IEEE Transactions on Automatic Control 43:1612–1617
7. de Vahl Davis G. (1986) Numerical Methods in Engineering and Science. Allen & Unwin, London
8. Ferretti G., Magnani G., Rocco P. (1998) Modeling, Identification, and Compensation of Pulsating Torque in Permanent Magnet AC Motors. IEEE Transactions on Industrial Electronics 45:912–920
9. Freyermuth B. (1991) An Approach to Model Based Fault Diagnosis of Industrial Robots. Proc. of 1991 IEEE International Conference on Robotics and Automation, Sacramento, CA, 1350–1356
10. Frank P.M. (1990) Fault Diagnosis in Dynamic Systems Using Analytical and Knowledge-Based Redudancy – A Survey and Some New Results. Automatica 26:459–474
11. Frank P.M. (1996) Analytical and Qualitative Model-Based Fault Diagnosis – A Survey and Some New Results. European Journal of Control 2:6–28
12. Gertler J. (1998) Fault Detection and Diagnosis in Engineering Systems. Marcel Dekker Inc., New York
13. Isermann R., Freyermuth B. (1991) Process Fault Diagnosis Based on Process Model Knowledge – Part I: Principles for Fault Diagnosis with Parameter Estimation. ASME J. of Dynamic Systems, Measurement, and Control 113:620–626
14. Isermann R., Freyermuth B. (1991) Process Fault Diagnosis Based on Process Model Knowledge – Part II: Case Study, Experiment. ASME J. of Dynamic Systems, Measurement, and Control 113:627–633
15. Neumann C.P., Tourassis V.D. (1985) Discrete Dynamic Robot Models. IEEE Transactions on Systems, Man, and Cybernetics 15:193–204
16. Patton R.J., Frank P.M., Clark R.N. (2000) Issues in Fault Diagnosis for Dynamic Systems. Springer-Verlag, London

17. Patton R.J., Uppal F.J., Lopez-Toribio C.J. (2000) Soft Computing Approaches to Fault Diagnosis for Dynamic Systems: A Survey. Preprints of the 4th IFAC Symposium on Fault Detection Supervision and Safety for Technical Processes, Budapest, H, 298-311
18. Polycarpou M.M., Helmicki A.J. (1995) Automated Fault Detection and Accomodation: A Learning Systems Approach. IEEE Transactions on Systems, Man, and Cybernetics 25:1447–1458
19. Schneider H., Frank P.M. (1996) Observer-Based Supervision and Fault Detection in Robots Using Nonlinear and fuzzy logic residual evaluation. IEEE Transactions on Control Systems Technology 4:274–282
20. Sciavicco L., Siciliano B. (2000) Modeling and Control of Robot Manipulators. Springer-Verlag, London
21. Takahashi R.H.C., Peres P.L.D. (1999) Unknown Input Observers for Uncertain Systems: A Unifying Approach. European Journal of Control 5:261–275
22. Trunov A.B., Polycarpou M.M. (2000) Automated Fault Diagnosis in Nonlinear Multivariable Systems Using a Learning Methodology. IEEE Transactions on Neural Networks 11:91–101
23. Vemuri A.T. (2001) Sensor Bias Fault Diagnosis in a Class of Nonlinear Systems. IEEE Transactions on Automatic Control 46:949–954
24. Vemuri A., Polycarpou M.M. (1996) Robust Nonlinear Fault Diagnosis in Input–Output Systems. International Journal of Control 68:343–360
25. Vemuri A.T., Polycarpou M.M., Diakourtis S.A. (1998) Neural Network Based Fault Detection in Robotic Manipulators. IEEE Transactions on Automatic Control 14:342–348
26. Visinsky M.L., Cavallaro J.R., Walker I.D. (1995) A Dynamic Fault Tolerance Framework for Remote Robots. IEEE Transactions on Robotics and Automation 11:477–490
27. Youcef-Toumi K., Ito O. (1990) A Time Delay Controller Design for Systems with Unknown Dynamics. ASME Journal of Dynamic Sytems, Measurement, and Control 112:133–142

A Survey of Fault Detection/Tolerance Strategies for AUVs and ROVs

Gianluca Antonelli

Dipartimento di Automazione, Elettromagnetismo,
Ingegneria dell'Informazione e Matematica Industriale
Università degli Studi di Cassino
via G. Di Biasio 43, 03043 Cassino (FR), Italy

Abstract. The use of Remotely Operated Vehicles (ROVs) and Autonomous Underwater Vehicles (AUVs) increased significantly in the last years. Such vehicles are complex systems engaged in missions in un-structured, unsafe environments for which the degree of autonomy becomes a crucial issue. In this sense, the capability to detect and tolerate faults is a key to successfully terminate the mission or recuperate the vehicle. In this paper, an overview of fault detection and fault tolerance algorithms, specifically designed for ROVs or for AUVs is presented.

1 Introduction

Autonomous Underwater Vehicles (AUVs) and Remotely Operated Vehicles (ROVs) received increasing attention in the last years due to their significant impact in several underwater operations. Examples are the monitoring and maintenance of off-shore structures or pipelines, or the exploration of the sea bottom; see, e.g., reference [50] for a complete overview of existing AUVs with description of their possible applications and the main subsystems. The benefit in the use of unmanned vehicles is in terms of safety, due to the possibility to avoid the risk of manned missions, and economic. Generally, AUVs are required to operate over long periods of time in unstructured environments in which an undetected failure usually implies loss of the vehicle. It is clear that, even in case of failure detection, in order to terminate the mission, or simply to recover the vehicle, a fault tolerant strategy, in a wide sense, must be implemented. In fact, simple system failure can cause mission abort [28] while the adoption of a fault tolerant strategy allows to safely terminate the task as in the case of the arctic mission of Theseus [18]. In case of the use of ROVs, a skilled human operator is in charge of command the vehicle, a failure detection strategy is then of help in the human decision making process. Based on the information detected, the operator can decide in the vehicle rescue or to terminate the mission by, e.g., turning off a thruster.

Fault detection is the process of monitoring a system in order to recognize the presence of a failure; fault isolation or diagnosis is the capability to determine which specific subsystem is subject to failure. Often in literature there is a certain overlapping in the use of these terms. Fault tolerance is the

F. Caccavale, L. Villani (Eds.): Fault Diagnosis and Fault Tolerance for Mechatronic Systems, STAR 1, pp. 109-127, 2003.

capability to complete the mission also in case of failure of one or more subsystems, it is referred also as fault control, fault accommodation or control reconfiguration. In the following the terms fault detection/tolerance will be used.

The characteristics of a fault detection scheme are the capability of isolate the detected failure; the sensitivity, in terms of magnitude of the failure that can be detected and the robustness in the sense of the capability of working properly also in non-nominal conditions. The requirements of a fault tolerant scheme are the reliability, the maintainability and survivability [40]. The common concept is that, to overcome the loss of capability due to a failure, a kind of redundancy is required in the system. A general scheme is presented in Figure 1.

In this paper, a survey over existing fault detection and fault tolerant schemes for underwater vehicles is presented. For these specific systems, adopting proper strategies, an hardware-software (hw-sw) sensor failure or an hw-sw thruster failure can be successfully handled in different operating conditions as it will be shown in next Sections. In some conditions, it is required that the fault detection scheme is also able to diagnostic some external not-nominal working conditions such as a multi-path phenomena affecting the echo-sounder system [11]. It is worth noticing that, for autonomous systems such as AUVs, space systems or aircraft, a fault tolerant strategy is necessary to safely recover the damaged vehicle and, obviously, there is no *panic button* in the sense that the choice of turning off the power or activate some kind of brakes is not available.

Most of the fault detection schemes are model-based [1–4,9,11,20,22,39,48] and concern the dynamic relationship between actuators and vehicle behavior or the specific input-output thruster dynamics. A model-free method is presented in [7,29]. Higher level fault detection schemes are presented in [17,18,25,51]. References [6,30,32,33,44,49] deal with hardware/software aspects of a fault detection implementation for AUVs. Neural Network and Learning techniques have also been presented [15,16,21,23,46].

Concerning fault tolerant schemes, most of them consider a thruster redundant vehicle that, after a fault occurred in one of the thrusters, still is actuated in 6 degrees of freedom (dofs). Based on this assumption a reallocation of the desired forces on the vehicle over the working thrusters is performed [2–4,9,11,35–38,41,48]. Of interest is also the study of reconfiguration strategies if the vehicle becomes under-actuated [34].

Only few papers concern the experimental results of fault detection and fault tolerant schemes [2–4,6,9,11,17,18,32,35,36,41,48]; for all the above references it is worth noticing that the successfully results has been achieved with the implementation of simple algorithms.

In Section 2, the mathematical model of underwater vehicles is briefly discussed and the main properties are highlighted; in Section 3 a small list of failures occurred during wet operations is reported; Section 4 and 5 report the description of fault detection and tolerant strategies for underwater vehicles. Since the implementation of such strategies in a real environment is not trivial Section 6 describes in more detail some successfully experiments. Finally, the conclusions are drawn in Section 7.

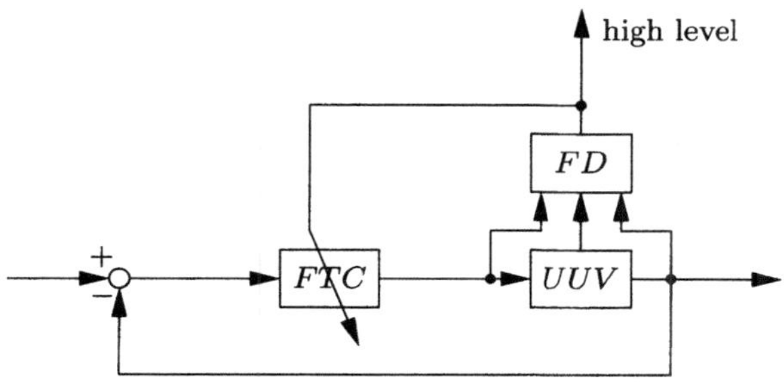

Fig. 1. General fault detection/tolerant control scheme for an Unmanned Underwater Vehicle (UUV). The Fault Detection (FD) block is in charge of detecting the failure, send a message to the higher level supervisor and, eventually, modifying the Fault Tolerant Controller (FTC).

2 Modeling

Let define as $\Sigma_i, \{O - \boldsymbol{xyz}\}$ a reference frame that we will suppose earth-fixed and inertial and as $\Sigma_v, \{O_v - \boldsymbol{x}_v \boldsymbol{y}_v \boldsymbol{z}_v\}$ a vehicle-fixed frame. The unit vector $\boldsymbol{z}$ is considered parallel to the gravity, $\boldsymbol{x}_v$ is parallel to the vehicle fore aft direction and $\boldsymbol{z}_v$ is aligned with $\boldsymbol{z}$ when the vehicle is at the surface. Moreover, $\boldsymbol{\eta} = \begin{bmatrix} \boldsymbol{\eta}_1^{\mathrm{T}} & \boldsymbol{\eta}_2^{\mathrm{T}} \end{bmatrix}^{\mathrm{T}}$, where $\boldsymbol{\eta}_1 = \begin{bmatrix} x & y & z \end{bmatrix}^{\mathrm{T}} \in \mathrm{I\!R}^3$ is the vector of vehicle position coordinates in a earth-fixed reference frame, $\boldsymbol{\eta}_2 = \begin{bmatrix} \phi & \theta & \psi \end{bmatrix}^{\mathrm{T}} \in \mathrm{I\!R}^3$ is the vector of Euler-angles coordinates expressing the vehicle orientation in the earth-fixed reference frame. $\boldsymbol{\nu} = \begin{bmatrix} \boldsymbol{\nu}_1^{\mathrm{T}} & \boldsymbol{\nu}_2^{\mathrm{T}} \end{bmatrix}^{\mathrm{T}}$, $\boldsymbol{\nu}_1 \in \mathrm{I\!R}^3$ is the vector of vehicle linear velocity expressed in the vehicle-fixed reference frame, $\boldsymbol{\nu}_2 \in \mathrm{I\!R}^3$ is the vector of vehicle angular velocity expressed in the vehicle-fixed reference frame.

The vehicle-fixed velocity $\boldsymbol{\nu}$ and the time derivative of the earth-fixed vehicle coordinates are related by the following:

$$\boldsymbol{\nu}_1 = \boldsymbol{R}_I^B \dot{\boldsymbol{\eta}}_1, \tag{1}$$

$$\boldsymbol{\nu}_2 = \boldsymbol{T}(\boldsymbol{\eta}_2) \dot{\boldsymbol{\eta}}_2. \tag{2}$$

where $\boldsymbol{R}_I^B$ is the rotation matrix expressing the transformation from the earth-fixed frame to the vehicle-fixed frame, the matrix $\boldsymbol{T}(\boldsymbol{\eta}_2) \in \mathrm{I\!R}^{3\times 3}$, expressed in terms of Euler angles, is given, e.g., in [19].

The equations of motion of an AUV can be written in vehicle-fixed reference frame in the form [19]:

$$\boldsymbol{M}\dot{\boldsymbol{\nu}} + \boldsymbol{C}(\boldsymbol{\nu})\boldsymbol{\nu} + \boldsymbol{D}(\boldsymbol{\nu})\boldsymbol{\nu} + \boldsymbol{g}(\boldsymbol{R}_I^B) = \boldsymbol{\tau} \tag{3}$$

where $\boldsymbol{M} \in \mathbb{R}^{6\times 6}$ is the mass matrix including the added mass, $\boldsymbol{C}(\boldsymbol{\nu})\boldsymbol{\nu} \in \mathbb{R}^6$ is the vector of Coriolis and Centripetal terms including the effects of the added mass, $\boldsymbol{D}(\boldsymbol{\nu})\boldsymbol{\nu} \in \mathbb{R}^6$ is the vector of friction and hydrodynamic damping terms, $\boldsymbol{g}(\boldsymbol{R}_I^B) \in \mathbb{R}^6$ is the vector of gravitational and buoyant generalized forces, $\boldsymbol{\tau} \in \mathbb{R}^6$ is the vector of forces and moments acting on the vehicle.

For a ground-fixed serial chain of rigid bodies the properties of linearity in the dynamic parameters hold [43]. In case of underwater vehicles, adopting a suitable mathematical model for the hydrodynamic forces, eq. (3) can be rewritten in a matrix form that exhibits this property:

$$\boldsymbol{\Phi}(\boldsymbol{R}_I^B, \boldsymbol{\nu}, \dot{\boldsymbol{\nu}})\boldsymbol{\theta} = \boldsymbol{\tau} \tag{4}$$

with $\boldsymbol{\Phi} \in \mathbb{R}^{6\times n_\theta}$, being n_θ the size of $\boldsymbol{\theta}$, the vector of parameters. Notice that n_θ depends on the model used for the hydrodynamic generalized forces. An estimate is given in [24] where, due to the hydrodynamic terms, $n_\theta > 100$.

The dynamics of marine vehicles is affected by the ocean current. Let us assume that the ocean current, expressed in the earth-fixed frame, $\boldsymbol{\nu}_c^I$ is constant and irrotational, i.e., $\boldsymbol{\nu}_c^I = [\,\nu_{c,x} \quad \nu_{c,y} \quad \nu_{c,z} \quad 0 \quad 0 \quad 0]^{\mathrm{T}}$ and $\dot{\boldsymbol{\nu}}_c^I = \mathbf{0}$. Its effects can be added to the dynamics of a rigid body moving in a fluid simply considering the *relative* velocity in vehicle-fixed frame $\boldsymbol{\nu}_r = \boldsymbol{\nu} - \boldsymbol{R}_I^B \boldsymbol{\nu}_c^I$ in the derivation of the Coriolis, centripetal and the damping terms in the equation (3).

Thrusters and control surfaces provide forces and moments on the vehicle according to a nonlinear relation. A simplified relationship can be expressed through the linear mapping [19]

$$\boldsymbol{\tau} = \boldsymbol{B}\boldsymbol{u} \tag{5}$$

where $\boldsymbol{B}$ is a $(6 \times p)$ matrix known as the Thruster Control Matrix (TCM) and $\boldsymbol{u}$ is the $(p \times 1)$ vector of control inputs. In case of 6-dofs AUVs it is $p \geq 6$. In case of underactuated AUVs p is generally $p = 3$ and it is obtained by one thruster and 2 control surfaces. As an example, ODIN (Figure 2) has the following TCM:

$$\boldsymbol{B} = \begin{bmatrix} * & * & * & * & 0 & 0 & 0 & 0 \\ * & * & * & * & 0 & 0 & 0 & 0 \\ 0 & 0 & 0 & 0 & * & * & * & * \\ 0 & 0 & 0 & 0 & * & * & * & * \\ 0 & 0 & 0 & 0 & * & * & * & * \\ * & * & * & * & 0 & 0 & 0 & 0 \end{bmatrix} \tag{6}$$

where $*$ means a non-zero constant factor depending on the thruster allocation. Different TCM can be observed as in, e.g., the vehicle Phantom S3 manufactured by Deepocean Engineering that has 4 thrusters:

$$\boldsymbol{B} = \begin{bmatrix} * & * & 0 & 0 \\ 0 & 0 & * & * \\ 0 & 0 & * & * \\ 0 & 0 & * & * \\ * & * & 0 & 0 \\ * & * & 0 & 0 \end{bmatrix} \tag{7}$$

in which it can be recognized that not all the directions are independently actuated.

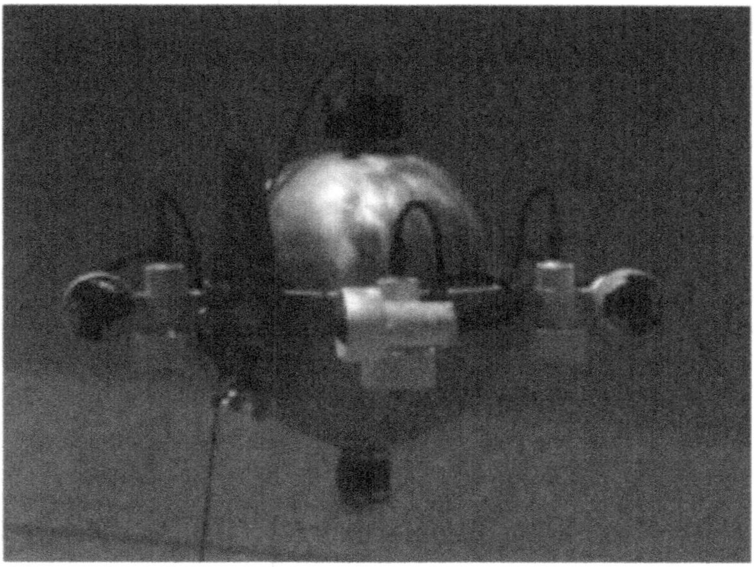

Fig. 2. ODIN AUV (Autonomous Systems Laboratory of the University of Hawaii, USA).

The brief description of the mathematical model clearly shows that the equations of motion are strongly non-linear and coupled. Moreover, the number of dynamic parameters is large; the hydrodynamic terms are the result of a mathematical approximation and their values are difficult to identify. The relationship in eq. (5) is an approximation that is not valid for control surfaces whose dynamic contribution is velocity-dependent.

3 Experienced Failures

In this section, a small list of possible ROVs/AUVs' failures is reported.

Sensor failure. The underwater vehicles are currently equipped with several sensors in order to provide information about their localization and velocity. The problem is not easy, it does not exist a single, reliable sensor that gives the required position/velocity measurement or information about the environment, e.g., about the presence of obstacles. For this reason the use of sensor fusion by, e.g., a Kalman filtering approach, is a common technique to provide to the controller the required variables. This structural redundancy can be used to provide fault detection capabilities to the system. In detail, a failure can occurs in one of the following sensors:

- IMU (Inertial Measurement Unit): provide information about the vehicle's linear acceleration and angular velocity;
- Depth Sensor: by measuring the water pressure gives the vehicle's depth;
- Altitude and frontal sonar: they are used to detect the presence of obstacles and the distance from the sea bottom;
- Ground Speed Sonar: it measures the linear velocity of the vehicle with respect to the ground;
- Currentmeter: it measures the relative velocity between vehicle and water;
- GPS (Global Positioning System): it is used to reset the drift error of the IMU and localize exactly the vehicle; it works only at the surface;
- Compass: it gives the vehicle yaw;
- Baseline Acoustic: with the help of one or more transmitters allows exact localization of the vehicle in a specific range of underwater environment;
- Vision system: it can be used to track structures such as pipelines.

For each of the above sensors the failure can consist in a output zeroing if, e.g., there is an electrical trouble or in an loss of meaning. It can be considered as sensor failure also an external disturbance such as a multi-path reading of the sonar that can be interpreted as a sensor fault and correspondingly detected.

Thruster blocking. It occurs when a solid body is between the propeller blades. It can be checked monitoring the current required by the thruster. It has been observed, e.g., during the Antarctic mission of Romeo [11] caused by a block of ice.

Flooded thruster. A thruster flooded with water has been observed during a Romeo's mission [11]. The consequence has been an electrical dispersion causing an increasing blade rotation velocity and thus a thruster force higher then the desired one.

Fin stuck or lost. This failure can causes a loss of steering capability as discussed by mean of simple numerical simulations in [22].

Rotor failure. A possible consequence of different failures of the thrusters is the zeroing of the blade rotation. The thruster in question, thus, simply stop working. This has been intentionally experienced during experiments with ODIN [35,36,48], RAUVER [20] and Roby 2 [3] and during another Romeo's mission [11].

Hardware-software failure. A crash in the hardware or software implemented on the vehicle can be experienced. In this case, redundancy techniques can be implemented to handle such situations [6].

Fig. 3. Romeo ROV (courtesy of CNR-IAN Robotlab).

4 Fault Detection Schemes

In [2,3] a model-based fault detection scheme is presented to isolate actuators' failures in the horizontal motion. Each thruster is modeled as in [19]. The algorithm is based on a bank of Extended Kalman Filters (EKFs) whose outputs are checked in order to detect behaviors not coherent with the dynamic model. In case of two horizontal thrusters and horizontal motion 3 EKFs are designed to simulate the 3 behaviors: nominal behavior, left thruster fault, right thruster fault. The cross-checking of the output allows efficient detection as it has been extensively validated experimentally (detailed are given in Section 6). A sketch of this scheme is given in Figure 4 where $\boldsymbol{u}$ is the vector of thruster inputs and the vehicle yaw ψ is measured by mean of a compass. In [4], the same approach is investigated with the use of a sliding-mode observer instead of the EKF. The effectiveness of this approach is also discussed by means of experiments.

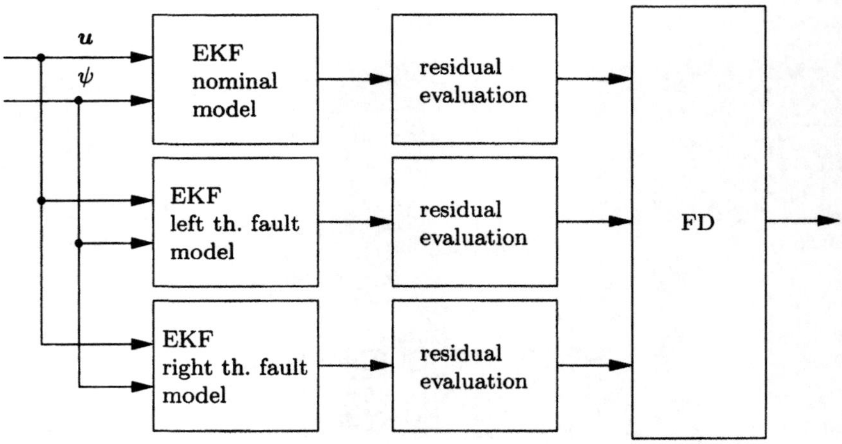

Fig. 4. Fault detection strategy for one of the horizontal thruster failure proposed by A. Alessandri, M. Caccia and G. Verrugio.

The work in [9,11] focuses on the thruster failure detection by monitoring the motor current and the propeller's revolution rate. The non-linear nominal characteristic has been experimentally identified, thus, if the measured couple current-propeller's rate is out of a specific bound a fault is experienced. Based on a mapping of the i-o axis the possible cause is also specified with a message to the remote human operator. The two failures corresponding to a thruster flooding or to a rotor failure, in fact, falls in different axis regions and can be isolated. Interesting experiments are given in Section 6.

The fault in a thruster is also monitored in [48] by the use of a hall-effect sensor mounted on all the thrusters. The input is the desired voltage as computed by the controller and the TCM, the output is the voltage as measured by the hall-effect sensor; the mismatching between the measured and the predicted voltage is considered as a fault. The paper also consider fault tolerance for sensor and actuator faults and experiments, as shown in next sections.

The vehicle Theseus [17,18] is equipped with a Fault Manager subsystem. This provides some kind of high-level failure detection in the sense that the mission is divided in a number of phases (each phase is a series of manoeuvres between waypoints); in case of failure of a phase there is a corresponding behavior to be activated. See Section 6 for detail about a practical intervention of the Fault Manager. A hierarchical control system developed for future implementation on Theseus is described in [51], this is based on the layered control concepts [10].

References [6,32] present an architecture for AUVs that integrates fault detection capabilities of the subsystems. The hardware and software architecture, named AUVC (Autonomous Underwater Vehicle Controller) implements a fault detection strategy based on five rule-based systems that monitor all the subsystems. The five systems concern the Navigation, the Power/Propulsion, the Direction Control and the Communication; they are coordinated by a Global Diagnoser that avoid contradictory actions. A specific attention has been given at the hardware reliability, in fact the AUVC is distributed on a redundant network of 18 loosely coupled processor. AUVC has been also used to test the approach proposed in [33], a redundancy management technique based on CLIPS expert system shell to identify faults affecting depth and heading control. In [49] an architecture developed for the vehicle ARICS with fault detection/tolerant capabilities is presented. A software developed for ROVs to help the remote operator that integrate some elementary fault detection algorithms is presented in [44]. The paper in [30] describes the first results on the development of a long endurance AUV that is currently ongoing at the MBARI (Monterey Bay Aquarium Research Institute, California, USA). The fault detection approach is mainly ported from the MIT (Massachusetts Institute of technology, Massachusetts, USA) vehicle Odyssey (I and II) [8] and it is based on the Layered Control Architecture. The software architecture is based on C++, QNX-based modules, that give multi-tasking capabilities suited for fault-tolerant operations. A single thread suppression and restart can be implemented to recover from software failure. Short-duration operations in open sea and long-duration operations in the lab proved the effectiveness of this approach.

In [22] a model-based observer is used to generate residual between the sensor measured behavior and the predicted one. The model also take into account the presence of waves in case of operations near the surface. When the residual is larger than a given threshold a Fuzzy Inference System is in charge of isolate the source of this mismatching (see sketch in Figure 5). A planar simulation is provided in the case of low speed under wave action and a stuck fin.

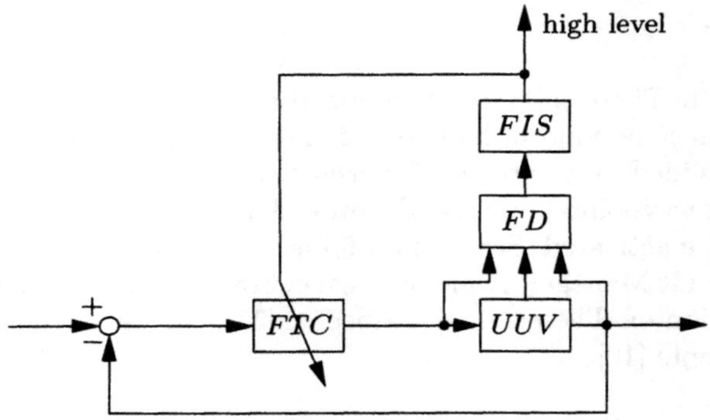

Fig. 5. Fuzzy fault detection/tolerant control scheme proposed by A.J. Healey: the FIS (Fuzzy Inference System) block is in charge of isolation of the faults observed by the FD (Fault Detection) block between fin stroke, servo error, residual and wave activity detectors.

A model-free fault detection method is proposed in [7,29], this is based on the Hotelling T^2 statistic and it is a data-driven approach. The validation is based on a 6-dofs simulation affected by stern plane jams and rudder jams.

A model-based, integrated heterogeneous knowledge approach is proposed in [20]. A multi-dimensional correlation analysis allows to increase the confidence in the detected fault and to detect also indirectly sensed subsystems. Some preliminary results with the vehicle RAUVER are also given.

In [1] a model-based fault detection scheme for thrusters and sensors is proposed. It has been designed based on the identified model of the 6-thruster ROV Linotip and it is composed on a bank of single-output Luenberger observers. Its effectiveness is verified by simulations. Another model-based fault detection scheme verified by simulations is provided in [39]. A robust approach in [31].

A neuro-symbolic hybrid system is used in [15] to perform fault diagnosis on AUVs with learning capability. The method is simulated on the planar motion of the VORTEX mathematical model. Another learning technique is proposed in [16] and verified by means of 6-dofs simulations. In [46], a neural network mathematical model is used to set-up a self-diagnosis scheme of the AUV. A software for health monitoring of AUVs' missions with learning capabilities is also described in [27].

The work in [5] studies a systematic, quantitative approach in order to maximize the mission and return success probabilities. The failed sensor is de-activated and the information obtained by a backup sensor able to recover the vehicle. No dynamic simulations are provided.

In [14], the wavelet theory is used to detect the fault in the vehicle's navigation angle fault.

Finally, [42] developed a software tool to test intelligent controllers for AUVs. This is done by using learning techniques from the artificial intelligence theory.

5 Fault Tolerant Schemes

Most of the fault tolerant controllers developed for thruster-driven underwater vehicles are based on a suitable inversion of the TCM in eq. (5). It is self-evident that, if the matrix is low rectangular it is still possible to turn off the broken thruster and to control the vehicle in all the 6 dofs. When the vehicle becomes underactuated, or when it is driven by control surfaces, the problem is mathematically more complex. In this case only few solutions to specific set-up have been developed.

The work in [9,11] reports a fault tolerant approach for ROMEO, a thruster redundant ROV with 8 thrusters. The strategy, experimentally verified, simply consists in deleting the column corresponding to the broken thruster from the TCM. The mapping from the vehicle force/moment to the thrusters' forces, thus, does not concern the broken component. A similar approach is used in [48] by exploiting the thruster redundancy of ODIN, an AUV developed at the Autonomous Systems Laboratory (ASL) of the University of Hawaii, HI, USA.

In [35–38,41], a task-space-based, fault tolerant control for vehicles with redundant actuation is proposed. The control law is model based and handle the thruster redundancy by a pseudo-inverse approach of the TCM that guarantee the minimization of the actuator quadratic norm. The thruster dynamics, with the model described in [26], is also taken into account. The proposed approach is sketched in Figure 6 where the subscript d denotes the desired trajectory, $\boldsymbol{V}_m$ is motor input voltage and Ω the propeller angular velocity of the thrusters. Despite the necessity to know the dynamic parameters the experimental validation was successfully (see Section 6).

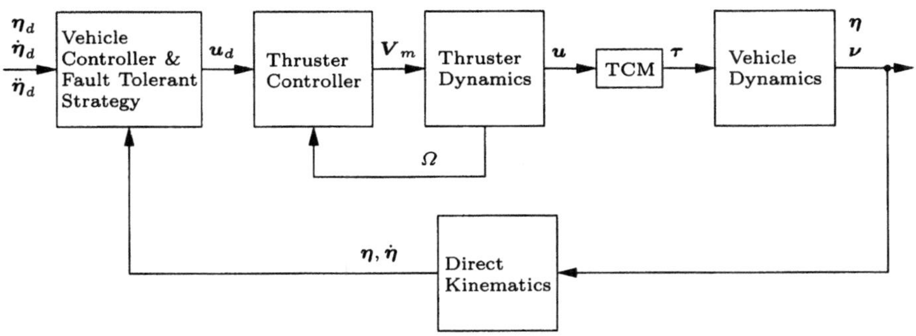

Fig. 6. Fault tolerance strategy proposed by N. Sarkar and T.K. Podder.

The work [48] presents a fault detection-tolerant scheme for sensor faults. In detail, the depth of the vehicle is measured by using a pressure sensor and a bottom sonar sensor. A third, virtual, sensor is added, this is basically an ARX (AutoRegressive eXogeneous) model of the vehicle depth dynamics. By comparing the measured values with the predicted ones, the residual is calculated and the failed sensor eventually disconnected for the remaining mission. A sketch of the scheme is shown in Figure 7, in nominal working condition the 3 residual R_i are close to the null value. It is worth noticing that this approach require exact knowledge of the ocean depth.

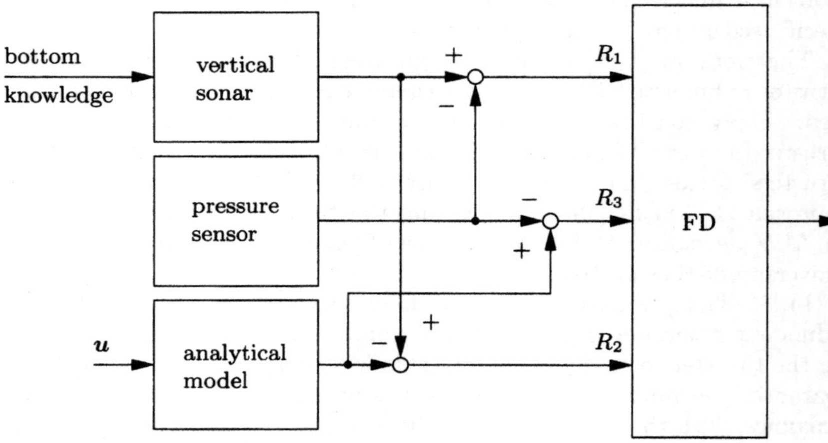

Fig. 7. Fault tolerance strategy for sensor fault proposed by K.C. Yang, Y. Yuh and S.K. Choi.

In [34], the case of an underactuated AUV controlled by control surfaces is considered. The vehicle tries to move in the unactuated dofs by using elementary motions in the actuated dofs. The method has been tested on the vehicle ARCS and showed that, in this form, it is not applicable. This study gives information on structural changes to adopt in the vehicle in order to develop a vehicle suitable for the implementation of this method.

One of the first works of reconfiguration control for AUVs is given in [45], where, however, only a superficial description of a possible fault tolerant strategy is provided. Recently, [47] proposed a reconfiguration strategy to accommodate actuator faults, this is based on a mixed H_2/H_∞ problem. Simulation results are provided.

6 Experiments

Roby 2 is a ROV developed at the Naval Automation Institute (CNR-IAN), National Research Council, Italy. It has been object of several wet tests to validate fault detection approaches. The horizontal motion is obtained by the use of two fore thrusters that control the surge velocity as well as the vehicle heading; the depth is regulated thanks to two vertical thrusters. In [2–4] experiments of different fault detection schemes has been carried out by causing an actuator failure: one of the thrusters has been simply turned off. The experiments in [2,3] have been carried out in a pool, in [4] the experiments also concern a comparison between EKFs and sliding-mode observers. The latter is a result of a bilateral project with the Naval Postgraduate School, Monterey, CA.

The Italian Naval Automation Institute (CNR-IAN) also developed the ROV ROMEO and tested, in an antarctic mission, both fault detection and tolerant schemes [9,11]. In particular the case of flooded and blocked thrusters occurred. In both cases the fault has been detected and the information could have been reported to the human operator to activate the reconfiguration procedure. Figure 8 shows the expected and measured motor currents in case of flooded thruster, it can be observed a persistent mismatching between the output of the model and the measured values.

The vehicle Theseus manufactured by ISE Research Ltd with the Canadian Department of National Defence successfully handled a failure during an Arctic mission of cable laying [17,18]. In detail, the vehicle did not terminate an homing step, probably due to poor acoustic conditions, and the Fault Manager activated a safe behavior: stop under the ice and wait for further instructions. This allowed to re-establish acoustic telemetry and surface tracking and to safely recover the vehicle. Notice that his fault wasn't intentionally caused [18].

The vehicle ODIN, an AUV developed at the Autonomous Systems Laboratory (ASL) of the University of Hawaii, HI, USA, has been used for several experiments. In [48] the fault detection and tolerant schemes are experimentally validated. The thruster fault has been tested by zeroing the output voltage by software, the fault detection scheme identified the trouble and correctly reconfigured the force allocation by proper modification of the TCM. The fault tolerant scheme with respect to depth sensor fault has also been tested by zeroing the sensor reading and verifying that the algorithm, after a programmed time of 1 s, correctly switched on the other sensor. While the theory has been developed for a 6-dofs vehicle, the experiments results only present the vehicle depth.

The same vehicle has been used to validate the fault tolerant approach developed in [35,36,41] in 6-dofs experiments. Different experiments have been carried out by zeroing the voltage on one or two thrusters simultaneously that, however, did not make the vehicle became under-actuated. The implemented control law is based on an identified reduced ODIN model and does not make

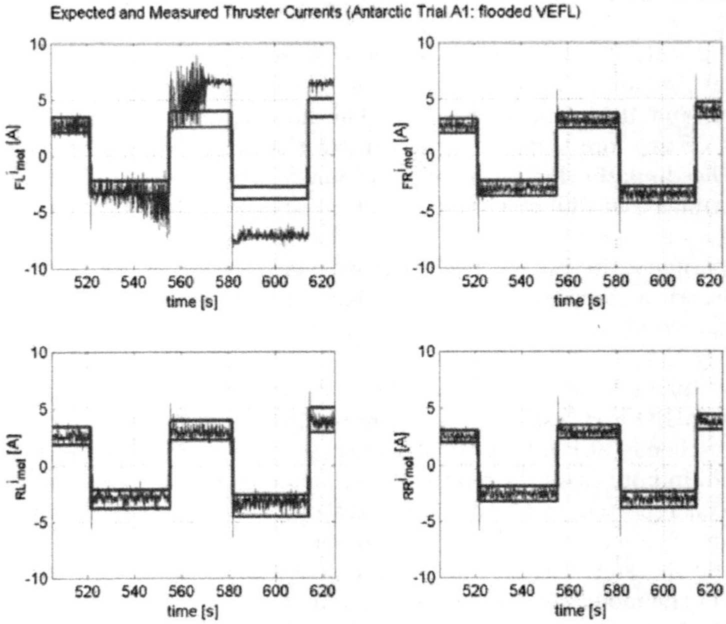

Fig. 8. Expected and measured motor currents for the vehicle Romeo in case of flooded thruster (courtesy of CNR-IAN Robotlab).

use of thruster model neither it needs the vehicle acceleration as required by the theory; the block diagram, thus, is simply given by Figure 9. Details on the control law are given in the referenced papers, the basic formulation of the controller is given by:

$$\boldsymbol{u} = \boldsymbol{E}^{\dagger}\left[(\ddot{\boldsymbol{\eta}}_d - \boldsymbol{\beta}) + \boldsymbol{K}_v\dot{\tilde{\boldsymbol{\eta}}} + \boldsymbol{K}_p\tilde{\boldsymbol{\eta}}\right] \tag{8}$$

where $\boldsymbol{K}_v$ and $\boldsymbol{K}_v$ are control gains, the vector $\tilde{\boldsymbol{\eta}}$ is the position/orientation error, $\boldsymbol{\beta}$ represents the compensation of the nonlinear terms of the equation of motions. The matrix $\boldsymbol{E}$ takes into account the TCM matrix, the inertia matrix and the Jacobian matrix that convert body-fixed to inertial-fixed velocities. Generalization about control of a desired task is given in [36]. The experiments validated the proposed approach; in Figure 10 the voltages are shown, it can be recognized that thrusters 2 and 6 (one horizontal and one vertical) are turned off at $t = 260$ s and $t = 300$ s, respectively, this causes an augmentation of the chattering of the remaining thrusters that, however, still can perform the desired task (Figure 11).

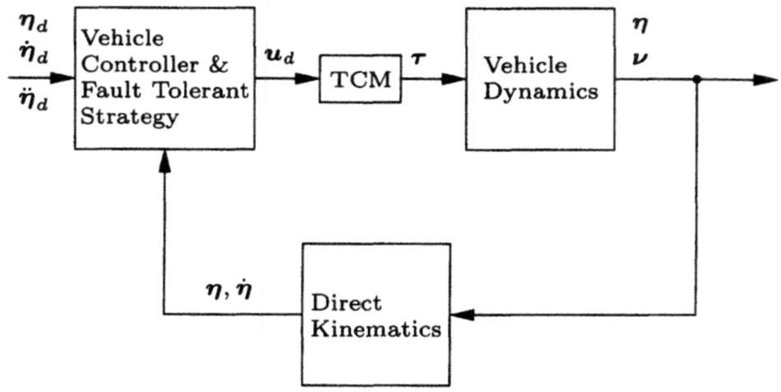

Fig. 9. Sketch of the fault tolerance strategy implemented by N. Sarkar, T.K. Podder and G. Antonelli.

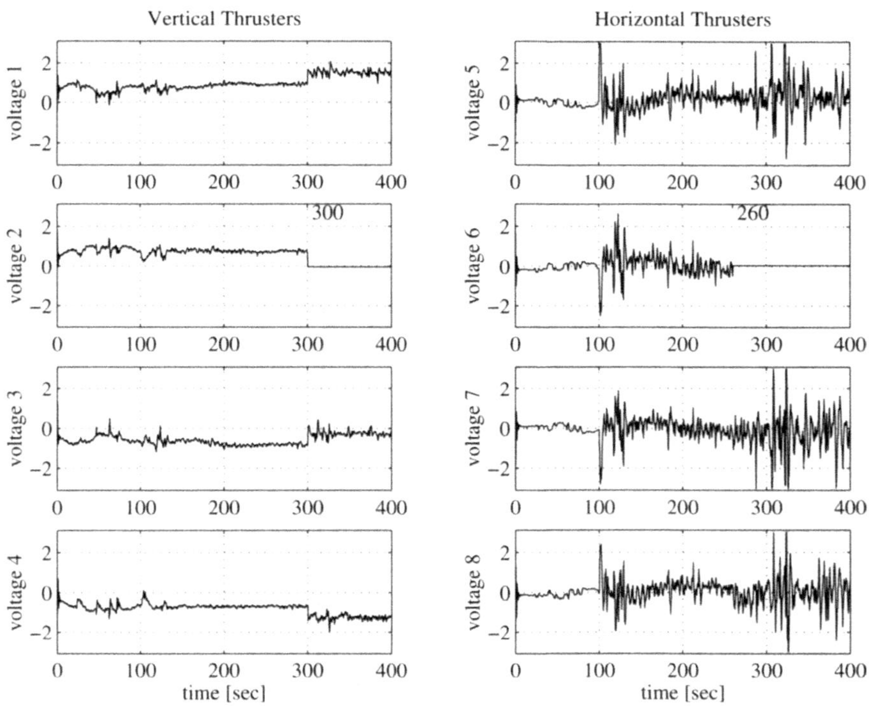

Fig. 10. Voltage profile - vertical thrusters (on the left) and horizontal thrusters (on the right) for the N. Sarkar and T.K. Podder algorithm.

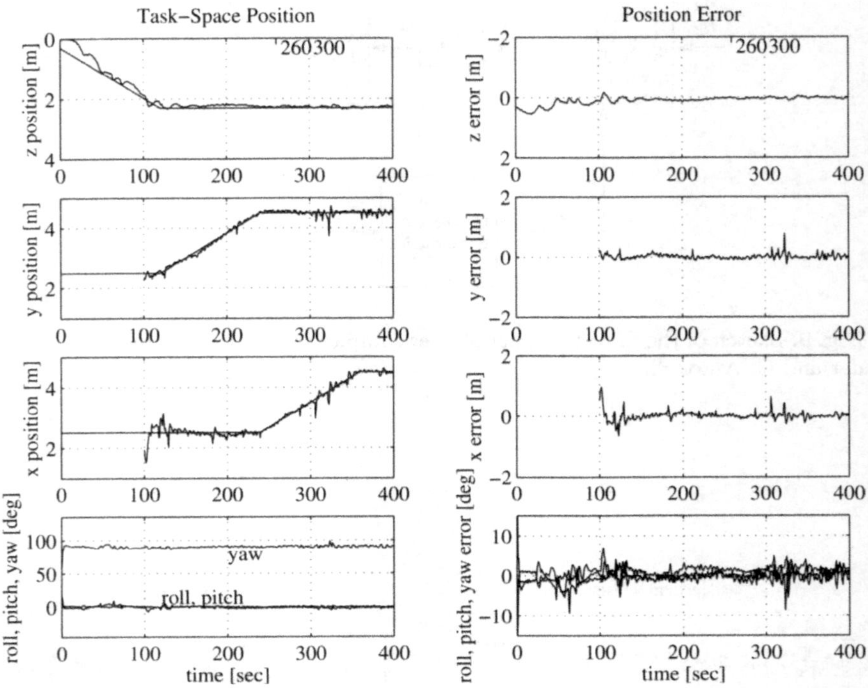

Fig. 11. Task-space trajectories (left column) and their errors (right column) for the N. Sarkar and T.K. Podder algorithm.

The AUVC described in [6,32] has been tested on a six-processor version on the Large Diameter Unmanned Underwater Vehicle of the Naval Undersea Warfare Center.

7 Conclusions

An overview over existing fault detection and fault tolerant schemes for underwater vehicles has been presented. The case of failures for autonomous missions in un-structured environment is, obviously, a dramatic occurrence to handle. In this sense, the underwater community would benefit from research studies with a strong practical orientation rather than theoretical-only approaches. Failures in a redundant sensor seem to be a solved problem; however, care have to be done in the tuning of the detection gains, real-data experiments for off-line tuning seems to be a reliable way to select those gains. As far as thrusters are of concern, experiments have shown that current AUVs can be controlled at low velocity with 6 thrusters, also if the original symmetric allocation is lost, without a strong performance deterioration. Some possible research areas concern the case of a thruster-driven AUV that become under-actuated and the case of vehicles controlled by means of control surfaces that failed; in both cases some practical approaches might be useful for the underwater community.

References

1. Alekseev Y.K., Kostenko V.V., Shumsky A.Y. (1994) Use of Identification and Fault Diagnostic Methods for Underwater Robotics. Oceans'94, Brest, FR, 489–494
2. Alessandri A., Caccia M., Verruggio G. (1998) A Model-Based Approach to Fault Diagnosis in Unmanned Underwater Vehicles. Oceans'98, Nice, F, 825–829
3. Alessandri A., Caccia M., Verruggio G. (1999) Fault Detection of Actuator Faults in Unmanned Underwater Vehicles. Control Engineering Practice 7:357–368
4. Alessandri A., Hawkinson T., Healey A.J., Veruggio G. (1999) Robust Model-Based Fault Diagnosis for Unmanned Underwater Vehicles Using Sliding Mode-Observers. Proc. Int. Symposium Unmanned Untethered Submersible Technology
5. Babcock IV P.S., Zinchuk J.J. (1990) Fault Tolerant Design Optimization: Application to an Autonomous Underwater Vechile Navigation System. Proc. Symposium on Autonomous Underwater Vehicle Technology, 34–43.
6. Barnett D., McClaran S. (1996) Architecture of the Texas A& M Autonomous Underwater Vehicle Controller. Proc. Symposium on Autonomous Underwater Vehicle Technology, 231–237
7. Beale G.O., Kim J.H. (2000) A Robust Approach to Reconfigurable Control. 5th IFAC Conference on Manoeuvring and Control of Marine Craft, Aalborg, DK, 197–202
8. Bellingham J.G., Goudey C.A., Consi T.R., Bales J.W., Atwood D.K., Leonard J.J., Chryssostomidis C. (1994) A second generation survey AUV. Proc. Symposium on Autonomous Underwater Vehicle Technology, 148–155
9. Bono R., Bruzzone Ga., Bruzzone Gi., Caccia M. (1999) ROV Actuator Fault Diagnosis Through Servo-Amplifiers' Monitoring: an Operational Experience. MTS/IEEE Oceans'99, Seattle, WH, 3:1318–1324
10. Brooks R.A. (1986) A Robust Layered Control System for a Mobile Robot. IEEE Jour. of Robotics and Automation 2:14–23
11. Caccia M., Bono R., Bruzzone Ga., Bruzzone Gi., Spirandelli E., Verruggio G. (2001) Experiences on Actuator Fault Detection, Diagnosis and Accomodation for ROVs. Proc. Int. Symposium Unmanned Untethered Submersible Technology, Durham, NH
12. Caccia M., Verruggio G. (2000) Guidance and Control of a Reconfigurable Unmanned Underwater Vehicle. Control Engineering Practice 8:21–37
13. Cancilliere F.M. (1994) Advanced UUV Technology. Oceans'94, Brest, FR, 147–151
14. Demin X., Lei G. (2000) Wavelet Transform and its Application to Autonomous Underwater Vehicle Control System Fault Detection. Proc. 2000 Int. Symposium Underwater Technology, 99–104

15. Deuker B., Perrier M., Amy B. (1998) Fault-Diagnosis of Subsea Robots Using Neuro-Symbolic Hybrid Systems. Oceans'98, Nice, F, 830–834
16. Farrell J., Berger T., Appleby B.D. (1993) Using Learning Techniques to Accomodate Unanticipated Faults. Control Systems Magazine 13:(3)40–49
17. Ferguson J.S. (1998) The Theseus Autonomous Underwater Vehicle. Two Successful Missions. Proc. 1998 Int. Symposium Underwater Technology, 109–114
18. Ferguson J.S., Pope A., Butler B., Verrall R. (1999) Theseus AUV - Two Record Breaking Missions. Sea Technology Magazine, 65–70
19. Fossen T. (1994) Guidance and Control of Ocean Vehicles. John Wiley & Sons, Chichester, UK
20. Hamilton K., Lane D., Taylor N., Brown K. (2001) Fault Diagnosis on Autonomous Robotic Vehicles with RECOVERY: an Integrated Heterogeneous-Knowledge Approach. IEEE Int. Conference on Robotics and Automation, San Francisco, CA, 1251–1256
21. Healey A.J. (1992) A Neural Network Approach to Failure Diagnostics for Underwater Vehicles. Proc. IEEE Oc. Eng. Society Symposium on Autonomous Underwater Vehicles, Washington D.C.
22. Healey A.J. (1998) Analytical Redundancy and Fuzzy Inference in AUV Fault Detection and Compensation. Proc. Oceanology 1998, Brighton, 45–50
23. Healey A.J., Bahrke F., Navarrete J. (1992) Failure Diagnostics for Underwater Vehicles: A Neural Network Approach. IFAC Conference on Maneuvering and Control of Marine Craft, 293–306
24. Healey A.J., Lienard D. (1993) Multivariable Sliding Mode Control for Autonomous Diving and Steering of Unmanned Underwater Vehicles. IEEE Journal of Oceanic Engineering 18:327–339
25. Healey A.J., Marco D.B. (1992) Experimental Verification of Mission Planning by Autonomous Mission Execution and Data Visualization Using the NPS AUV II. Proc. 1992 Symposium Autonomous Underwater Vehicle Technology, 65–72
26. Healey A.J., Rock S.M., Cody S., Miles D., Brown J.P. (1995) Toward an Improved Understanding of Thruster Dynamics for Underwater Vehicles. IEEE Journal of Oceanic Engineering 20:354–361
27. Hornfeld W., Frenzel E. (1998) Intelligent AUV On-Board Health Monitoring Software (INDOS). Oceans'98, Nice, F, 815–819
28. Hutchison B. (1991) Velocity Aided Inertial Navigation. Proc. Sensor Nav. Issues for UUVs. C.S. Draper Lab
29. Kim J.H., Beale G.O. (2001) Fault Detection and Classification in Underwater Vehicle Using the T^2 Statistic. 9th Mediterranean Conference on Control and Automation, Dubrovnik, Kr
30. Kirkwood W.J., Shane F., Gashler D., Au D., Thomas H., Sibenac M., O'Reilly T.C., Konvalina T., McEwen R., Bahlavouni A., Tervalon N., Bellingham J.G. (2001) Development of a Long Endurance Autonomous Underwater Vehicle for Ocean Science Exploration. MTS/IEEE Conference and Exhibition Oceans 2001,1504–1512
31. Mangoubi R.S., Appleby B.D., Verghese G.C., VanderVelde W.E. (1995) A Robust Failure Detection and Isolation Algorithm. Proc. 34th Conference Decision & Control, New Orleans, LA, 2377–2382
32. Nelson E., McClaran S., and Barnett D. (1996) Development and Validation of the Texas A& M University Autonomous Underwater Vehicle Controller. Proc. Symposium on Autonomous Underwater Vehicle Technology, 203–208

33. Orrick A., McDermott M., Barnett D., Nelson E., Williams G. (1994) Failure Detection in an Autonomous Underwater Vehicle. Proc. Symposium on Autonomous Underwater Vehicle Technology, 377–382
34. Perrault D., Nahon M. (1998) Fault-Tolerant Control of an Autonomous Underwater Vehicle. Oceans'98, Nice, F, 820–824
35. Podder T.K., Antonelli G., Sarkar N. (2000) Fault Tolerant Control of an Autonomous Underwater Vehicle Under Thruster Redundancy: Simulations and Experiments. IEEE Int. Conference on Robotics and Automation, San Francisco, CA, 1251–1256
36. Podder T.K., Antonelli G., Sarkar N. (2001) An Experimental Investigation into the Fault-Tolerant Control of an Autonomous Underwater Vehicle. Journal of Advanced Robotics 15:501–520
37. Podder T.K., Sarkar N. (1999) Fault Tolerant Decomposition of Thruster Forces of an Autonomous Underwater Vehicle. IEEE Int. Conference on Robotics and Automation, Leuven, B, 84–89
38. Podder T.K. Sarkar N. (2001) Fault Tolerant Control of an Autonomous Underwater Vehicle Under Thruster Redundancy. Robotics and Autonomous Systems 34:39–52
39. Rae G.J.S., Dunn S.E. (1994) On-Line Damage Detection for Autonomous Underwater Vehicle," *Proc. Symposium on Autonomous Underwater Vehicle Technology, 383–392*
40. Rauch H.E. (1994) Intelligent Fault Diagnosis and Control Reconfiguration. IEEE Int. Symposium on Intelligent Control, Chicago, IL, 6–12
41. Sarkar N., Podder T.K., Antonelli G. (2002) Fault Accommodating Thruster Force Allocation of an AUV Considering Thruster Redundancy and Saturation. IEEE Transactions on Robotics and Automation 18:223–233
42. Schultz A.C. (1992) Adaptive Testing of Controllers for Autonomous Vehicles. Proc. Symposium on Autonomous Underwater Vehicle Technology, 158–164
43. Sciavicco L., Siciliano B. (2000) Modeling and Control of Robot Manipulators, Springer-Verlag, London, UK
44. Stokey R.P. (1994) Software Design Technique for the Man Machine Interface to a Complex Underwater Vehicle. Oceans'94, Brest, FR, 119–124
45. Tacconi G., Tiano A. (2001) Reconfigurable Control of an Autonomous Underwater Vehicle. Proc. Int. Symposium Unmanned Untethered Submersible Technology, 486–493.
46. Takai M., Fujii T., Ura T. (1995) A Model Based Diagnosis System for Autonomous Underwater Vehicles using Artifical Neural Networks. Proc. Int. Symposium Unmanned Untethered Submersible Technology, Durham, NH, 243–252
47. Tong G., Jimao Z. (1998) A Rapid Reconfiguration Strategy for UUV Control. Proc. 1998 Int. Symposium Underwater Technology, 478–483
48. Yang K.C., Yuh J., Choi S.K. (1998) Experimental Study of Fault-Tolerant System Design for Underwater Robots. IEEE Int. Conference on Robotics and Automation, Leuven, B, 1051–1056
49. Yavnai A. (1996) Architecture for an Autonomous Reconfigureable Intelligent Control System (ARICS). Proc. Symposium on Autonomous Underwater Vehicle Technology, 238–245
50. Yuh J., West M. (2001) Underwater Robotics. Journal of Advanced Robotics 15:609–639
51. Zheng X. (1992) Layered Control of a Practical AUV. Proc. Symposium on Autonomous Underwater Vehicle Technology, 142–147

Failure Detection, Identification and Reconfiguration in Flight Control

Jovan D. Bošković and Raman K. Mehra

Scientific Systems Company, Inc.
500 W. Cummings Park, Suite 3000
Woburn, MA 01801

Abstract. In this chapter we describe issues arising in Failure Detection, Identification and Reconfiguration (FDIR) in flight control. The fault accommodation problem is stated and discussed, followed by the derivation of models of sensor and control effector failures and structural damage. Corresponding FDIR schemes that minimize the effect of different failures and structural damage are presented next, and their properties are illustrated through numerical simulations of advanced combat aircraft. At the end, an Integrated FDIR scheme is discussed consisting of on-line health monitoring and status determination system combined with the proposed FDIR algorithms.

1 Introduction

The next generation of complex multivariable systems such as aircraft, spacecraft, autonomous vehicles, missiles, submarines and others, will be required to achieve and maintain the desired levels of performance under large perturbations and to perform multiple tasks in multiple operating regimes autonomously and under different failures and upsets. The overall closed-loop system will also be expected to be robust to modeling errors, noise, and different uncertainties. Since the dynamics of such systems are highly nonlinear and interacting and may often possess nonminimum phase characteristics, the corresponding control design problem is highly complex. The complexity of the problem is further increased by the requirement that the resulting control system should also be designed to take into account hard constraints on the actuators and the structural limits of the system.

On-line Health Monitoring and Failure Detection, Identification and Reconfiguration (HM-FDIR) has been widely applied in many different areas of technology including process control, power system control, manufacturing, and aerospace. In this context many different techniques have been developed and successfully implemented. The common goal shared by the designers is to achieve very high levels of safety and reliability even while minimizing operation and production costs. Fast and accurate FDI is of utmost importance since timely FDI information enables the operator (or the control computer) to take appropriate actions and minimize the effect of the failures and other upsets.

F. Caccavale, L. Villani (Eds.): Fault Diagnosis and Fault Tolerance for Mechatronic Systems, STAR 1, pp. 129-167, 2003.

In this context, from the point of view of HM-FDIR, modern commercial and military aircraft have some unique features compared to other industrial applications, and the design of control systems capable of compensating for a wide range of failures and upsets poses numerous challenges for the flight control designer. This is described in more detail in the following section.

1.1 Unique Challenges in Aircraft FDIR Design

There is a number of features, particularly in the case of advanced combat aircraft and Unmanned Combat Aerial Vehicles (UCAV), that distinguish FDIR in flight control from other industrial applications. Those include:
1. Modern aircraft are equipped by thrust vectoring and unconventional control surfaces, resulting in an over-actuated system where the number of control effectors exceeds the number of controlled variables. Thrust vectoring refers to the use of engines with moving jets that can create forces in different directions, while unconventional control surfaces include different flaps that are not commonly used for feedback control.
2. Most of the relevant aircraft states are measured and available for feedback. While commonly measured states include pitch, roll and yaw rates and angles, forward velocity and altitude, modern aircraft are equipped by sensors that also measure angle-of-attack and side-slip angle. Accelerations of several variables are also available, either through dedicated sensors, or through filtering.
3. Modern aircraft are characterized by highly reliable components and flight software. Even though individual sensors and control effectors are highly reliable by design, modern aircraft are equipped by triple-redundant control systems, and (at least) doubly-redundant flight computers. Flight software is flight-certified through rigorous Verification and Validation (V&V) procedures involving a very large number of computer simulations of different flight regimes encountered throughout the flight envelope.
4. Critical subsystem or component failures can cause instability of the closed-loop control system and loss of aircraft. While not every failure is critical, there is a large number of flight-critical failures (or their combinations) in every flight regime that can lead to substantial performance deterioration and instability. The fact that even a single failure may lead to catastrophic consequences makes the FDIR system design highly challenging.
5. One of the unique situations encountered in military aircraft is that due to flying over enemy territory when the aircraft is exposed to enemy fire from the air and ground. This may result in severe structural damage and/or failures and upsets whose effect may be such that the recovery is not possible. Another type of upset that is likely to occur during the flight over enemy territory are deteriorated communications due to jamming that can also result in performance degradation.

1.2 Existing Results in Aircraft FDIR

In the past decade there has been substantial progress in the development of on-line FDIR techniques in aerospace applications [1,3,2,13,14,9,8,11,6,5,7]. A large number of techniques has been proposed, and some of those have actually been flight tested [3]. Extensive simulation studies and flight-tests have demonstrated the potential of on-line FDIR systems to achieve the desired flight performance despite severe flight-critical subsystem or component failures, structural damage and large external disturbances.

While most of the techniques are based on treating the failures and structural damage as parametric uncertainty, and designing corresponding adaptive controllers to achieve the desired control objective, different methods were shown to be able to deal with very different size of the corresponding uncertainty sets. For instance, standard approach to indirect adaptive control is based on "certainty equivalence" where uncertain plant parameters are estimated on-line, and those estimates are in turn used in the control law assuring the stability of the overall system. However, it was shown that such an approach, when applied to linearized aircraft models, can handle only small to moderate uncertainty, which is due to a large number of model parameters that need to be estimated on-line. In the context of fault-tolerant control, this implies that standard indirect adaptive control approach can handle a very limited class of subsystem or component failures. It was also shown in [3,13,14] that direct adaptive control using neural networks can handle moderate to large uncertainty arising due to different failures and structural damage. However, it has yet to be demonstrated that such a method is well suited for combined sensor and control effector failures and structural damage that generate very large internal disturbances.

In addition, a systematic approach to the design of Integrated FDIR systems is lacking in the existing literature. The main precondition for effective FDIR design is a detailed analysis of different failure parameterizations, along with stability, robustness, and performance analysis of the overall closed-loop reconfigurable control system. In addition, integrated FDIR systems that can efficiently compensate for the *simultaneous* effect of sensor and control effector failures and structural damage, are not addressed adequately by existing methods. If such failures are not compensated for properly, this can lead to severe flight performance degradations, and even to the loss of aircraft. Hence integrated FDIR strategies that also take into account the highly coupled nonlinear aircraft dynamics in the nominal and failure modes are of great interest in practice.

To handle very large failure-generated uncertainties arising due to different types of failures and structural damage, we have developed a systematic procedure for the design of the corresponding FDIR systems. The procedure is based on the Multiple Models, Switching and Tuning (MMST) methodology from [20] described in the following section.

1.3 Multiple Models, Switching, and Tuning (MMST)

As shown in [12], for an aircraft controlled by a full-state feedback controller, lock-in-place failure introduces both constant and state-dependent disturbances into the overall closed-loop system, and control reconfiguration is necessary for maintaining the stability and robustness of the system. It was also shown that adaptive control using a single model may not be adequate for achieving this task in the presence of failures of critical control effectors. This is due to the fact that, in a particular flight regime, aircraft dynamics immediately after the failure may be very far from its nominal (no-failure) dynamics. Hence single model-based adaptive controller may be too slow to bring the closed-loop system close to the new operating regime, which may result in unacceptably large transients.

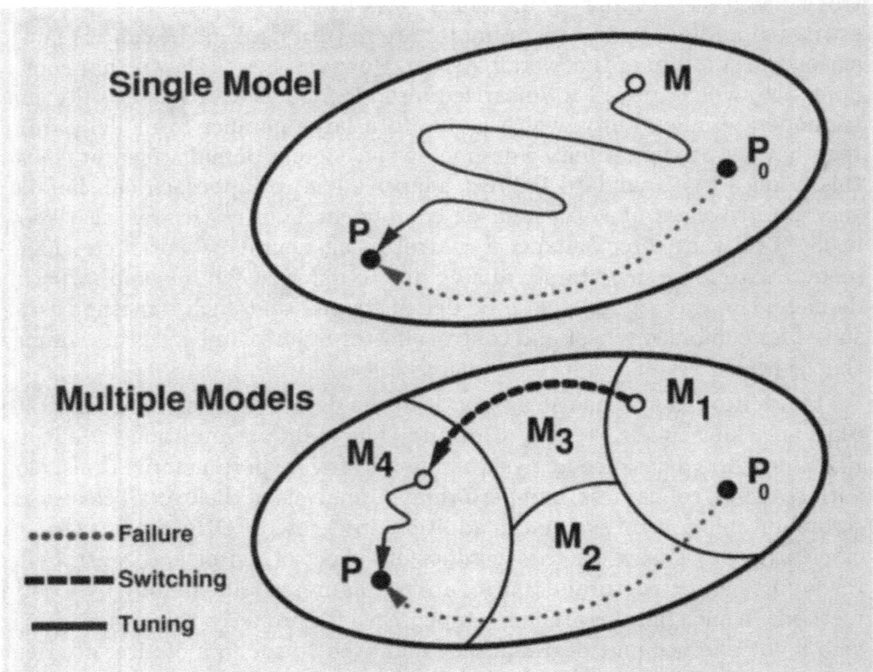

Fig. 1. *Single Model versus Multiple Model Adaptation: (i) (Top) Failure causes the plant to switch from some nominal set of parameters P_o to the failure set of parameters P; single-model adaptation may take a long time to identify the new set of parameters; (ii) (Bottom) The parametric set is divided into subsets; models M_i $(i = 1, 2, 3, 4)$ are placed into each subset; the multiple-model system switches to the model closest in some sense to the failure dynamics, and adapts from there thus achieving fast and accurate FDI (©1999-2002* Scientific Systems Company, Inc.*)*

For this reason, we have developed a multiple model-based reconfigurable control strategy based on the concept of Multiple Models, Switching and Tuning (MMST) from [20], Figure 2.

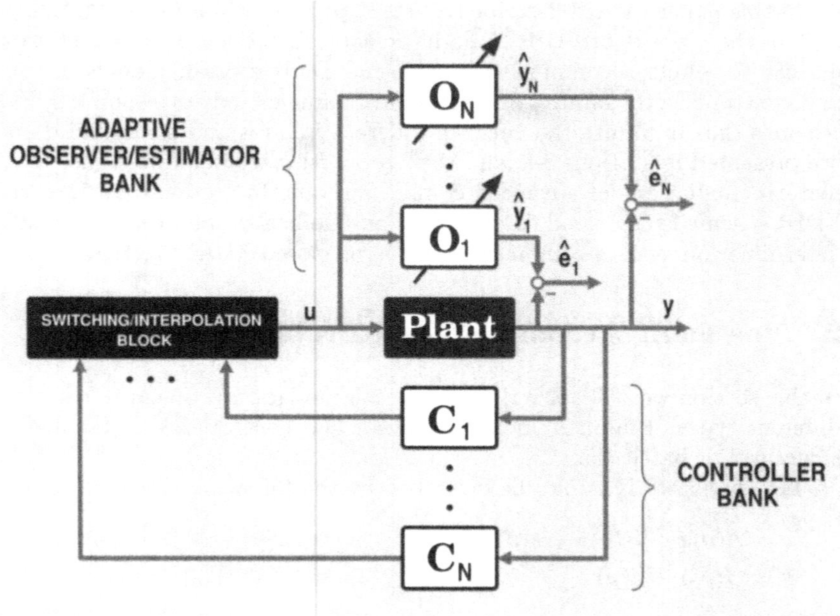

Fig. 2. *Structure of the Multiple Model-based Controller*

The MMST concept is fairly general and is well suited for the case of plants with rapidly varying dynamics. If the plant dynamics in different operating regimes is described by different models, the MMST concept is implemented by using the outputs of the parallel estimators (identification models) $M_1, M_2, \dots M_N$ to find the model closest in some sense to the current plant dynamics, and switch to the corresponding controller. In the context of reconfigurable control, each model in the above figure represents a different failure scenario. While the system is initiated with the controller for the no-failure case, in the case of failure the objective is to design a suitable control reconfiguration algorithm to assure that the scheme switches to the controller corresponding to the model closest in some sense to the dynamics of the failed plant. The need for MMST in plants with rapidly switching dynamics can be conveniently motivated using Figure 2.

As shown in the figure, failure may cause the plant dynamics to switch abruptly from some nominal point P_o in the parametric space, to the point P corresponding to the failed plant. The top figure illustrates the case when adaptation using a single model may be too slow to identify the new operating regime and reconfigure the controller. In such a case placing several models in the parametric set, switching to the model close to the dynamics of the failed plant, and adapting from there can result in fast and accurate control reconfiguration.

In this paper we will describe several implementations of the MMST concept in the context of FDIR in flight control. In section 2 we will state and discuss the fault accommodation problem. Derivation of models of sensor and control effector failures and structural damage and corresponding FDIR schemes that minimize the effect of different failures and structural damage are presented in sections 3-5, and their properties are illustrated through numerical simulations of advanced combat aircraft. In section 6 an Integrated FDIR scheme is discussed consisting of on-line health monitoring and status determination system combined with the proposed FDIR algorithms.

2 The Fault-Accommodation Problem

In this section we will state the fault-accommodation problem in the case of different types of failures, including sensor and control effector failures, and structural or battle damage.

Let the plant dynamics be described by the following model:

$$\dot{x} = A(p)x + B(p)u + z(p), \tag{1}$$

$$y = C(p)x + d(p) \tag{2}$$

where $x \in \mathbb{R}^n$, $u \in \mathbb{R}^m$ and $y \in \mathbb{R}^l$ denote respectively the state, control input, and output of the system, and p denotes a vector of failure-related parameters.

We assume that u is saturated, i.e. that $u \in \mathcal{S}_u = \{u : (u_i)_{min} \leq u_i \leq (u_i)_{max},\ (\bar{u}_i)_{min} \leq \dot{u}_i(t) \leq (\bar{u}_i)_{max},\ i = 1, 2, ..., m\}$. We also assume that $p \in \mathcal{S}_p$, where $\mathcal{S}_p$ denotes a failure set.

Nominal model: The above model is a fairly general representation of different failure conditions. For instance, in the no-failure case, the dynamics of the plant is described by

$$\dot{x} = A_o x + B_o u + z_o,$$

$$y = C_o x + d_o$$

where $A_o = A(p_N), B_o = B(p_N), z_o = z(p_N), d_o = d(p_N)$, and $C_o = C(p_N)$, where $p_N \in \mathcal{S}_p$ denotes the value of p that corresponds to the no-failure case. We next consider different failure models.

Sensor failure models: Common sensor failures include: (i) Bias; (ii) Drift; (iii) Performance degradation (loss of accuracy); (iv) Freezing; and (v) Calibration error (loss of effectiveness). In the case of freezing the output of the sensor is constant (but nonzero) regardless of the variations of the measured variable.

Sensor failures can be parameterized in the following way:

$$y_i(t) =$$

$$\begin{cases} x_i(t), & \text{for all } t \geq 0, \quad \textbf{No-Failure Case} \\ x_i(t) + d_i, \quad \dot{d}(t) \equiv 0, \quad d_i(t_{Fj}) \neq 0, & \textbf{Bias} \\ x_i(t) + d_i(t), \; |d_i(t)| = \lambda_i t, \; 0 < \lambda_i << 1, & \text{for all } t \geq t_{Fj}, \quad \textbf{Drift} \\ x_i(t) + d_i(t), \; |d_i(t)| \leq \bar{d}_i \; \dot{d}_i(t) \rightarrow 0, & \text{for all } t \geq t_{Fj}, \quad \textbf{Loss of Accuracy} \\ x_i(t_{Fi}), & \text{for all } t \geq k_{Fi}, \quad \textbf{Sensor Freezing} \\ c_i x_i(t), & \text{for all } t \geq k_{Fi}, \quad \textbf{Calibration Error} \end{cases}$$

where t_{Fi} denotes the time instant of failure of the ith sensor, and d_i denotes its accuracy coefficient such that $d_i \in [-\epsilon_{min}, \epsilon_{max}]$, where $\epsilon_{min}, \epsilon_{max} > 0$. Also, the effectiveness coefficient satisfies $c_i \in [\bar{c}_i, 1]$, where $\bar{c}_i > 0$. It is seen that all of the above failures can be modeled as changes in the matrix C and vector d in (2). These failures are illustrated in Figure 3. We can conclude that the model (2) includes all above cases, where C is a positive definite diagonal matrix whose elements vary slowly within $[\bar{c}_i, 1]$, and elements of vector d that vary slowly within $[-\bar{d}_i, \bar{d}_i]$ so that the following assumption is satisfied:

Assumption 1 $\dot{C}(t) \cong 0$; *and* $\dot{d}(t) \cong 0$.

Control Effector Failures: Typical control effector failures include: (i) Lock-In-Place (LIP); (ii) Hard-Over Failure (HOF); (iii) Float; and (iv) Loss of Effectiveness (LOE). In the case of LIP failures the effector "freezes" at a certain condition and does respond to subsequent commands. HOF is characterized by the effector moving to the upper or lower position limit regardless of the command. The speed of response is limited by the effector rate limit. Float failure occurs when the effector "floats" with zero moment and does not contribute to the control authority. Loss of effectiveness is characterized by lowering the effector gain with respect to its nominal value. Different types of control effector failures are shown in Figure 4.

Let us, for simplicity, assume that the transfer functions of the actuators is equal to one, and let $\tilde{u}$ be defined as:

$$\tilde{u} = Bu = BKu_c = [b_1 k_1 u_{c1} \quad b_2 k_2 u_{c2} \; \ldots \; b_m k_m u_{cm}], \tag{3}$$

where u_c denotes the vector of signals generated by the controller, $K = \text{diag}[k_1 \; k_2 \; \ldots \; k_m]$, and b_j denotes the jth column of B. The relationship between u_c and u is represented graphically in Figure 5. It is clear that in the case with no failures, $k_i = 1$, and $u_i = u_{ic}$, $i = 1, 2, \ldots, m$.

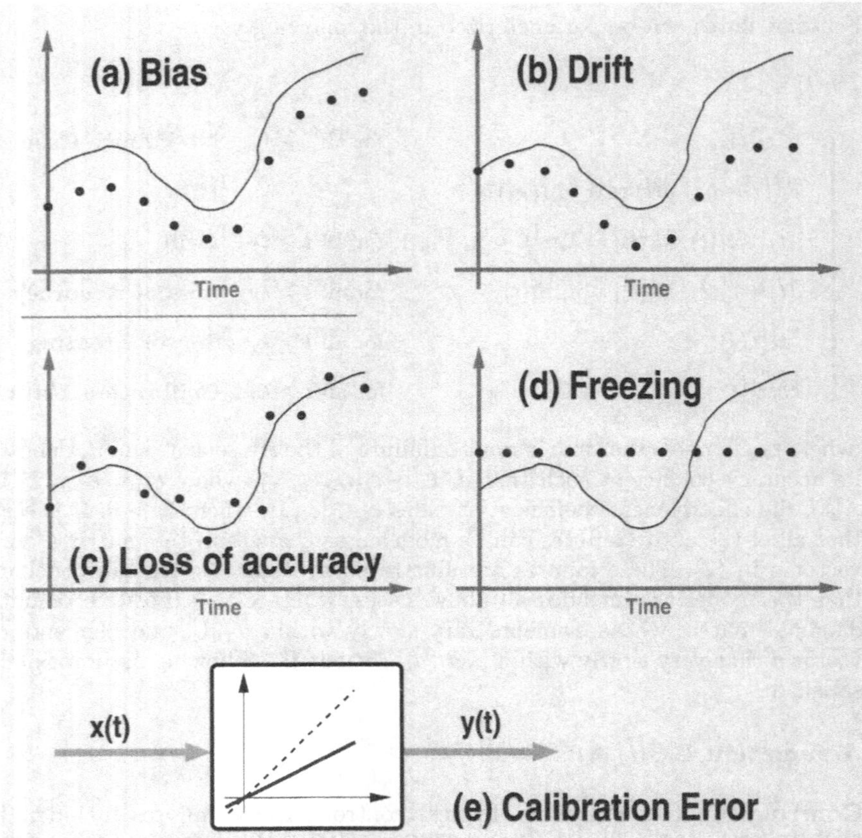

Fig. 3. *Types of Sensor Failures: (a) Bias; (b) Drift; (c) Performance degradation (loss of accuracy); (d) Freezing; and (e) Calibration error (loss of effectiveness)*

Different types of control effector failures can be parameterized as follows:

$$u_i(t) =$$

$$\begin{cases} u_c(t),\ k_i(t) = 1, & \text{for all } t \geq 0, & \textbf{No-Failure Case} \\ k_i(t)u_c(t),\ 0 < \epsilon_i \leq k_i(t) < 1, & \text{for all } t \geq t_{Fi}, & \textbf{Loss of Effectiveness} \\ 0,\ k_i(t) = 0, & \text{for all } t \geq t_{Fi}, & \textbf{Float Type of Failure} \\ u_{ci}(t_{Fi}),\ k_i(t) = 0, & \text{for all } t \geq t_{Fi}, & \textbf{Lock-in-Place Failure} \\ (u_i)_{min} \text{ or } (u_i)_{max},\ k_i(t) = 0, & \text{for all } t \geq t_{Fi}, & \textbf{Hard-Over Failure} \end{cases}$$

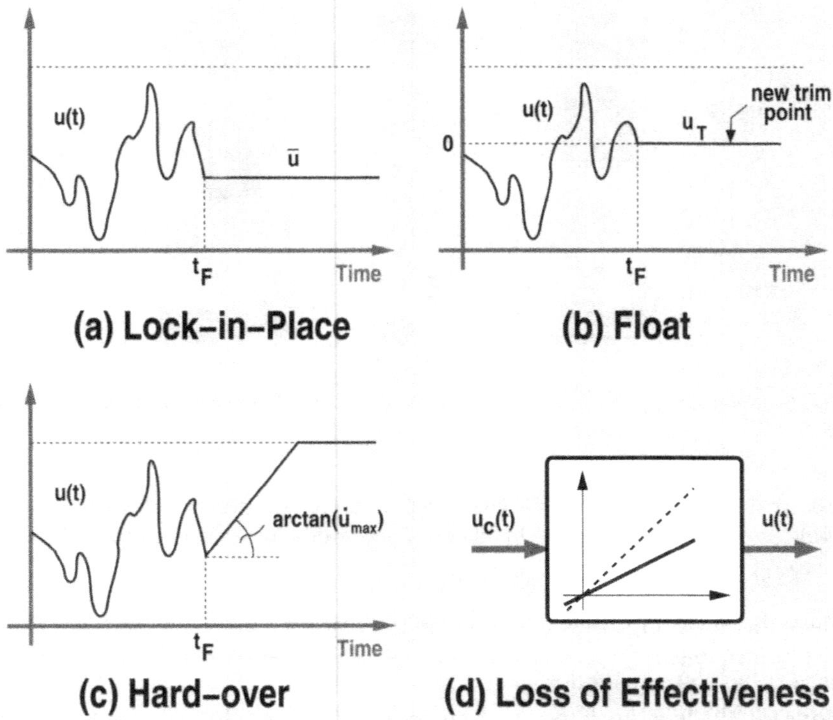

Fig. 4. *Types of Control Effector Failures: (a) Lock-in-place; (b) Float; (c) Hard-over; and (iv) Loss of effectiveness.*

where t_{Fi} denotes the time instant of failure of the ith effector, q_i denotes its effectiveness coefficient such that $q_i \in [\epsilon_i, 1]$, and $\epsilon_i > 0$ denotes its minimum effectiveness. Hence control effector failures can be modeled as changes in the matrix B and vector z in (2).

Structural or Battle Damage: Structural or battle damage can be modeled as simultaneous changes in matrices A and B [8]. This is discussed in detail in section 5.

Hence the model (1), (2) is a suitable representation covering a large class of failures and structural damage.

In order to present the failure accommodation approach, we will further simplify the model (1), (2) by introducing the following assumption:

Assumption 2 *(i) State $x(t)$ is measurable;*
(ii) $m > n$; and
(ii) $B(p)B^T(p)$ is full rank for all $p \in \mathcal{S}_p$.

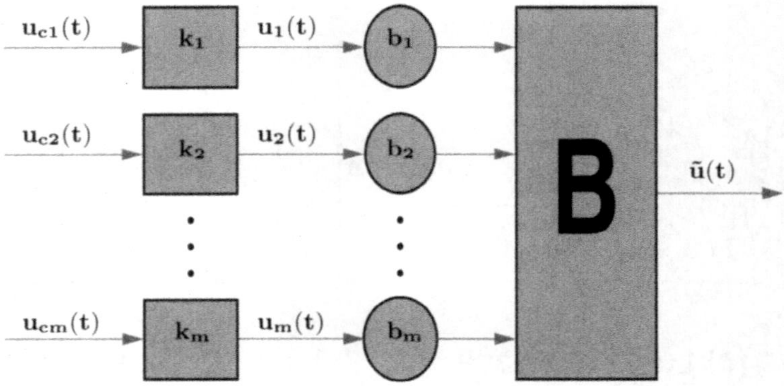

Fig. 5. *Relationship between u_c and u (*©1999-2002 Scientific Systems Company, Inc.*)*

Desired Dynamics: We further assume that the desired dynamics of the plant (1), (2) is defined by the state of the following reference model:

$$\dot{x}^*(t) = A_m x^*(t) + B_m r(t), \tag{4}$$

where A_m is asymptotically stable, and r denotes a vector of bounded piecewise continuous reference inputs. The control objective for the plant (1), (2) can now be stated as follows:

Control Objective: Design a control input $u(t)$ such that, for all $p \in \mathcal{S}_p$: (i) $\lim_{t\to\infty}[x(t) - x^*(t)] = 0$; and (ii) $u(t) \in \mathcal{S}_u$ for all time.

No-failure Model: Based on the above failure models, the model that describes plant dynamics in the no-failure and no-disturbance case is of the form:

$$\dot{x} = A_o x + B_o u,$$
$$y = x,$$

where A_o and B_o denote nominal system matrices.

Baseline Controller: We will choose a baseline controller for the no-failure model. We choose the following baseline control strategy, referred to as the Inverse Dynamics Control Law (IDCL):

$$u_c = W B_o^T (B_o W B_o^T)^{-1}\{-A_o x + A_m x + B_m r\}, \tag{5}$$

where u_c is the signal generated by the controller, and $W = W^T > 0$ denotes the control weighting matrix.

3 Sensor FDIR

In this paper, we will focus on the case when y is given by (2), and consider the following cases: (i) Case when $C = I$ and $d \neq 0$; (ii) Case when $C \neq I$ and

$d = 0$; and (iii) Case when $C \neq I$ and $d \neq 0$. The case (i) corresponds to the case of sensor drift, bias, and loss of accuracy, while the case (ii) corresponds to variations in the calibration gain. Since the sensor failures do not affect matrix B from (1), in this case $u_c = u$, i.e. the control signal generated by the controller coincides with that actually applied to the plant.

3.1 Sensor Drift, Bias, and Loss of Accuracy

The problem of sensor FDIR in this case reduces to estimating the unknown vector b that describes sensor drift, bias, and loss of accuracy. In this case we assume that $C = I$, so that the equation (2) is now of the form:

$$y = x + d. \tag{6}$$

We will also assume that $-\bar{d} \leq d \leq \bar{d}$ (elementwise), and $\bar{d}$ is a known vector.

We first take a derivative of y to obtain:

$$\dot{y} = \dot{x} = A(y - d) + Bu. \tag{7}$$

We now build a series-parallel estimator in the form:

$$\dot{\hat{y}} = A(y - \hat{d}) + Bu + \Lambda e_y, \tag{8}$$

where $e_y = \hat{y} - y$ and Λ is asymptotically stable. To simplify further analysis, we choose Λ to be diagonal.

The corresponding error equation is of the form:

$$\dot{e}_y = \Lambda e_y - A\phi_d, \tag{9}$$

where $\phi_d = \hat{d} - d$ denotes the parameter error vector. We next consider:

Theorem 1. *When $\hat{d}$ is adjusted using:*

$$\dot{\hat{d}} = \dot{\phi}_d = Proj_{[-\bar{d},\bar{d}]}\{\Gamma A^T e_y\}, \tag{10}$$

where Γ is a diagonal positive definite adaptive gain matrix, all signals in the system (9), (10) are bounded and $\lim_{t\to\infty} e_y(t) = 0$. In addition, if A is nonsingular, then $\lim_{t\to\infty} \phi_d(t) = 0$.

Proof: The proof of this theorem can be found in [10].

Tracking using the measurement estimate: In this case the control law is modified as:

$$u = W^{-1}B^T(BW^{-1}B^T)^{-1}\{-A(y - \hat{d}) - \Lambda e_y + A_m\hat{y} + B_m r\}. \tag{11}$$

This results in $\frac{d}{dt}(\hat{y} - x_m) = A_m(\hat{y} - x_m) + B_m r$. Hence $\lim_{t\to\infty}[\hat{y}(t) - x_m(t)] = 0$. Since $e_y(t) = \hat{y}(t) - y(t)$ also tends to zero asymptotically, we have that $y(t)$ will follow asymptotically the output of the reference model.

Static State Estimator: In some cases we are interested in assuring that the error $e_x(t) = x(t) - x_m(t)$, rather than $y(t) - x_m(t)$, tends to zero asymptotically. We will further address this problem. From (6) we have that $x = y - d$. We next build a state observer of the form:

$$\hat{x} = \hat{y} - \hat{d}.$$

The estimation error $e_x = \hat{x} - x$ is now:

$$e_x = e_y - \phi_d.$$

Hence, if A is nonsingular, $\lim_{t\to\infty} e_x(t) = 0$.
Tracking using state estimate: In this case the control law is modified as:

$$u = W^{-1}B^T(BW^{-1}B^T)^{-1}\{-(A - A_m)\hat{x} + B_m r\}. \tag{12}$$

Substituting the above expression into (2) and using (4), we obtain:

$$\dot{e}_c = A_m e_c - (A - A_m)\hat{e}_x, \tag{13}$$

where $e_c = x - x_m$ denotes the tracking error. Assuming that the estimator (9) is much faster than the system (13) and since $e_x(t)$ tends to zero asymptotically, we can conclude that $\lim_{t\to\infty} e_c(t) = 0$ as well.
Dynamic State Estimator: The stability of the system in the case of the static estimator for $x(t)$ depends on the speed of the estimator (refspe1). We will relax this assumption using a dynamic observer of the form:

$$\dot{\hat{x}} = A\hat{x} + Bu + L(\hat{y} - y),$$

where L is chosen to make $A + L$ asymptotically stable. The corresponding error model is obtained when (2) is subtracted from the above equation:

$$\dot{e}_x = Ae_x + L(\hat{y} - y), \tag{14}$$

where $e_x = \hat{x} - x$. Since the estimate of y is of the form $\hat{y} = \hat{x} + \hat{b}$, (14) can be rewritten as:

$$\dot{e}_x = (A + L)e_x + L\phi_d.$$

In the case of dynamic estimator, the adaptive law is chosen in the form:

$$\dot{\hat{d}} = \dot{\phi}_d = \text{Proj}_{[-\bar{d},\bar{d}]}\{-\Gamma e_y\}.$$

The stability of the resulting closed-loop system is discussed in [10].

3.2 Calibration Error

The problem of sensor FDIR in this case is a difficult one since the known signal (y) is a product of the unknown quantities C and x. There are two possible approaches that can be used. One is based on the estimation of the unknown C, and the other is based on estimating both C and $x(t)$. We will focus on the former case.

In this case we assume that $d = 0$, so that the equation (2) is now of the form:

$$y = Cx. \tag{15}$$

After taking a derivative of y, we obtain:

$$\dot{y} = C\dot{x} = CAC^{-1}y + CBu. \tag{16}$$

We next define $C_o = CAC^{-1}$. Since A and the bounds on elements of C are known, we can calculate the bounds on the elements of C_o. We next build a series-parallel estimator in the form:

$$\dot{\hat{y}} = \hat{C}_o y + \hat{C}Bu + \Lambda e_y, \tag{17}$$

where $e_y = \hat{y} - y$ and Λ is diagonal and asymptotically stable. The corresponding error equation is of the form:

$$\dot{e}_y = \Lambda e_y + \Phi\omega, \tag{18}$$

where $\Phi = \Theta - \Theta^*$, $\Theta = [\hat{C}_o^T \ \hat{C}^T]^T$, $\Theta^* = [C_o^T \ C^T]^T$, and $\omega = [y^T \ u^T B^T]^T$. We also denote the bounds on elements of Θ by a vector $\bar{\theta}$, and consider:

Theorem 2. *When $\Theta(t)$ is adjusted using:*

$$\dot{\Theta} = \dot{\Phi} = Proj_{[-\bar{\theta},\bar{\theta}]}\{-\Gamma\omega e_y^T\}, \tag{19}$$

where Γ is a diagonal positive definite adaptive gain matrix, all signals in the system (18), (19) are bounded and $\lim_{t\to\infty} e_y(t) = 0$. In addition, if ω is persistently exciting in the $n(n+1)$-dimensional parametric space corresponding to Θ, then $\lim_{t\to\infty} \Phi(t) = 0$.

Proof: The proof of this theorem can be found in [10].

Tracking using measurement estimate: In this case the control law is modified as:

$$u = \hat{C}^{-1}W^{-1}B^T(BW^{-1}B^T)^{-1}\{-\hat{C}_o y - \Lambda e_y + A_m\hat{y} + B_m r\}. \tag{20}$$

This results in $\frac{d}{dt}(\hat{y}-x_m) = A_m(\hat{y}-x_m)+B_m r$. Hence $\lim_{t\to\infty}[\hat{y}(t)-x_m(t)] = 0$. Since $e_y(t) = \hat{y}(t) - y(t)$ also tends to zero asymptotically, we have that $y(t)$ will follow asymptotically the reference model.

Static State Estimator: In this case we build a state observer of the form:

$$\hat{x} = \hat{C}^{-1}\hat{y}.$$

The estimation error $e_x = \hat{x} - x$ is now:

$$e_x = \hat{C}^{-1}e_y + (\hat{C}^{-1} - C^{-1})y.$$

Since $e_y(t)$ tends to zero asymptotically, $e_x(t)$ can tend to zero only if $\hat{C}(t) - C$ tends to zero, which will be so if ω is persistently exciting.
Tracking using state estimate: In this case the control law is modified as:

$$u = W^{-1}B^T(BW^{-1}B^T)^{-1}\{-(A - A_m)\hat{x} + B_m r\}. \tag{21}$$

Substituting the above expression into (2) and using (4), we obtain:

$$\dot{e}_c = A_m e_c - (A - A_m)\hat{e}_x, \tag{22}$$

where $e_c = x - x_m$ denotes the tracking error. Assuming again that the estimator (9) is much faster than the system (13) and since $e_x(t)$ tends to zero asymptotically, we can conclude that $\lim_{t\to\infty} e_c(t) = 0$ as well.

4 The Generalized Case

In this section we will now use the previous results to derive stable adaptive laws in the case when there is both the calibration error, and sensor bias or drift. In such a case we have:

$$y = Cx + d.$$

Upon taking the first derivative of the above equation, we obtain:

$$\dot{y} = C\dot{x} = CAC^{-1}(y - b) + CBu. \tag{23}$$

We also note that, based on the results from the previous section, we have that the following holds:

$$\begin{aligned} C_o(y-d) + CBu &= (\hat{C}_o - \Phi_o)(y - \hat{d} + \phi_d) + (\hat{C} - \Phi_c)Bu \\ &= \hat{C}A\hat{C}^{-1}(y - \hat{d}) + \hat{C}A\hat{C}^{-1}\phi_d + \hat{C}Bu \\ &\quad -\Omega(y - \hat{d}, u)\phi_c - \Phi_o\phi_d. \end{aligned}$$

Upon neglecting the term $\Phi_o\phi_d$ and using the notation $\hat{C}_o = \hat{C}A\hat{C}^{-1}$, we have that

$$\dot{y} = \hat{C}_o(y - \hat{d}) + \hat{C}_o\phi_d - \Omega(y - \hat{d}, u)\phi_c + \hat{C}Bu. \tag{24}$$

The observer is now chosen in the form:

$$\dot{\hat{y}} = \hat{C}_o(y - \hat{d}) + \hat{C}Bu + \Lambda e_y,$$

where Λ is asymptotically stable.

Hence the error model is now of the form:

$$\dot{e}_y = \Lambda e_y + \tilde{\Omega}\phi, \tag{25}$$

where $\tilde{\Omega} = [\hat{C}_o \;\; \Omega(y - \hat{d}, u)]$, and $\phi = [\phi_d^T \;\; \phi_c^T]^T$.

We now define $\hat{\theta} = [\hat{d}^T \;\; \hat{c}^T]^T$, and denote the bounds corresponding to the bounds on C and d as $[\theta_{min}, \theta_{max}]$. We next consider:

Theorem 3. *If $\hat{\theta}(t)$ is adjusted using:*

$$\dot{\hat{\theta}} = \dot{\phi} = Proj_{[\theta_{min},\theta_{max}]}\{-\Gamma\tilde{\Omega}^T e_y\}, \tag{26}$$

then $\lim_{t\to\infty} e_y(t) = \lim_{t\to\infty}[y(t) - x_m(t)] = 0$ *and all signals in the system (25),(26) are bounded. In addition, if $\tilde{\Omega}$ is persistently exciting in the $2n$-dimensional parametric space, then* $\lim_{t\to\infty}\phi(t) = 0$.

This theorem can be proved along the same lines as before. In this case the control law that assures that $\lim_{t\to\infty}[y(t) - x_m(t)] = 0$ is of the form:

$$u = \hat{C}^{-1}W^{-1}B^T(BW^{-1}B^T)^{-1}\{-\hat{C}A\hat{C}^{-1}(y - \hat{d}) + A_m\hat{y} + B_m r - \Lambda e_y\}.$$

Simulation 1*:* **Nominal response of F/A-18C/D during power approach.** *We first simulated the nominal (i.e. no-failure) case using the model from [11] of F/A-18C/D dynamics during power approach, and the baseline controller of the form (5). The control allocation algorithm is chosen to achieve the control objective without encountering position saturation. The resulting response is shown in Figure 6. Lateral response is omitted as all variables are essentially zero over the entire time interval. It is seen that the objective of following the states of the reference model (dashed line) is achieved with available control authority.*

Simulation 2*:* **Uncompensated sensor failures.** *We next simulated the aircraft response with the same controller, but now in the presence of bias and drift of angle-of-attack, pitch rate and pitch angle sensors. The vector d is assumed to change as:*

$$d_i(t) = e^{0.05*(t-t_F)}c_i, \ i = 2, 3, 4,$$

where $t_F = 3$*, and* $c = [0.0175\ 0.045\ \ -0.0525]^T$*. The response is shown in Figure 6. It is seen that the response is unacceptable and that effective failure compensation is needed in order to achieve the objective.*

Simulation 3*:* **Adaptive reconfigurable controller**. *We next simulated the aircraft response under the same failure scenario as in the previous simulation, but with the adaptive reconfigurable controller. The resulting response is shown in Figure 8, while the parameters and their estimates are shown in Figure 9. Controller adaptation is seen to result in substantially improved response.*

4.1 Conclusions

In this section we propose a suitable parametrization for the modeling of sensor failures in flight control applications. The failures include bias, drift, loss of accuracy, and calibration error. Based on this parameterization, reconfigurable control algorithms are designed and stability and robustness analysis is carried out. Properties of the proposed algorithms are illustrated through simulations.

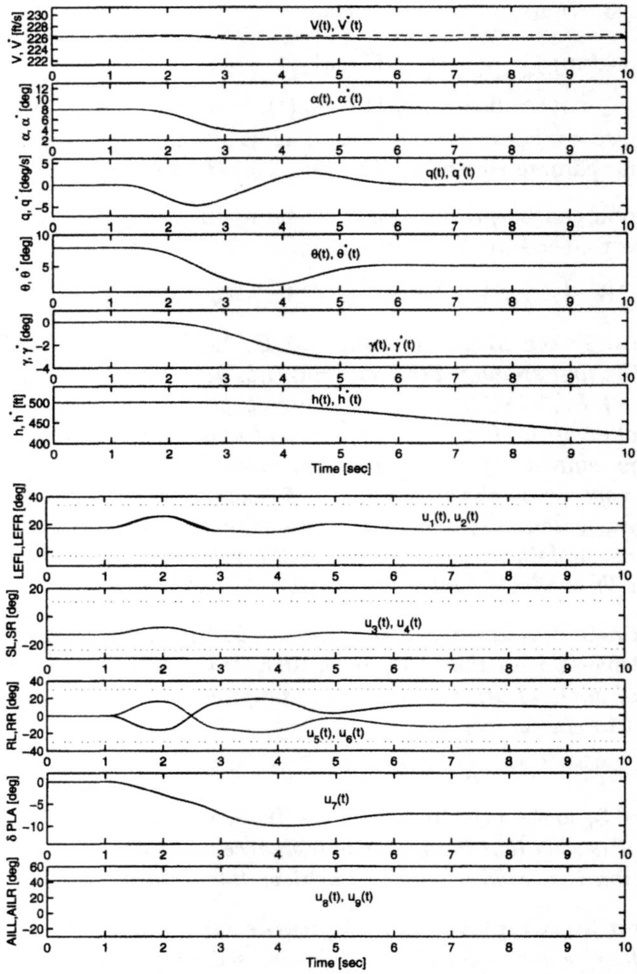

Fig. 6. *Simulated response of F/A-18C/D with the baseline controller in the no-failure case (solid line: actual response; dotted line: desired response; dashed lines denote position limits)*

5 Control Effector FDIR

In this section we present a new Failure Detection and Identification (FDI) and Adaptive Reconfigurable Control (ARC) scheme for achieving the desired flight performance in the presence of multiple control effector failures. The scheme is well suited for accommodation of failures that occur short time apart. The failures include lock-in-place and hard-over. The problem of stable control reconfiguration in the presence of such failures is solved using a scheme

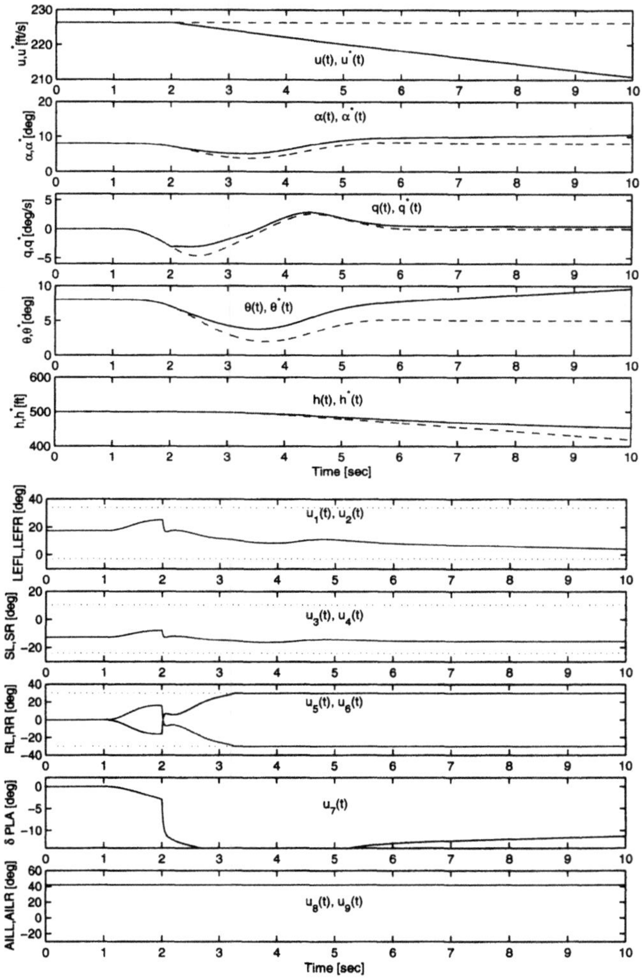

Fig. 7. *Aircraft response with the baseline controller in the case of bias and drift of angle-of-attack, pitch rate and pitch angle sensors (solid line: actual response; dotted line: desired response*

employing multiple adaptive FDI observers and controllers and a suitably chosen decision-making mechanism. The scheme is shown to guarantee the detection and identification of a failed effector. Due to a convenient structure of the overall system, the adaptive FDI observers can be modified on-line to reflect the new operating regime immediately following the failure. This enables fast FDI and effective ARC in the case of subsequent failures. The stability of the overall FDI-ARC system is demonstrated using the Lyapunov

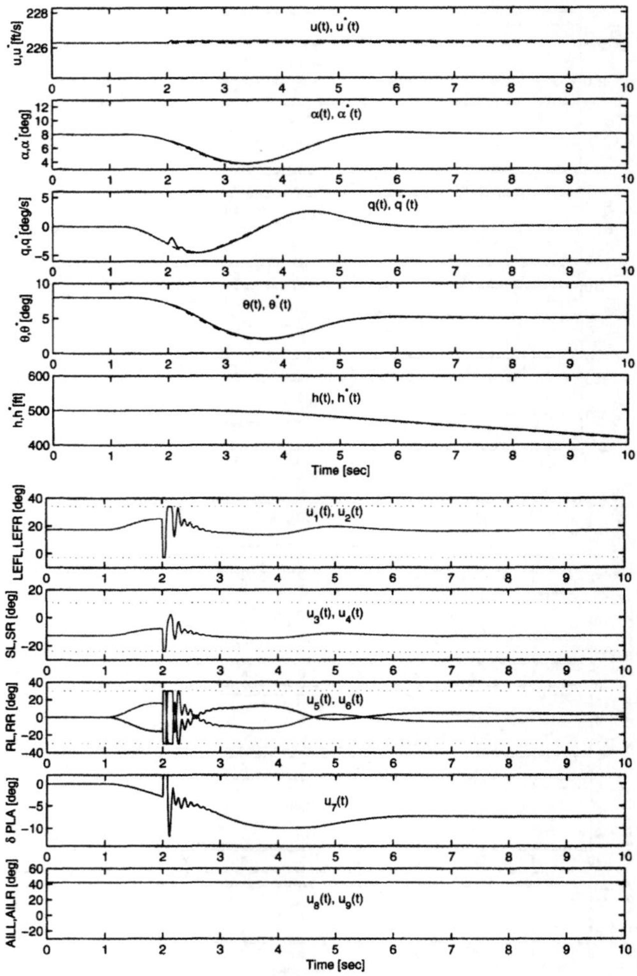

Fig. 8. *Aircraft response with the adaptive reconfigurable controller in the case of bias and drift of angle-of-attack, pitch rate and pitch angle sensors (solid line: actual response; dotted line: desired response)*

method, while the approach is illustrated through numerical simulations of the F-18 aircraft during carrier landing.

5.1 Failure Models

In this section we will focus on the case of lock-in-place and hard-over failures to derive FDI algorithms for guaranteed detection of these two types of failures. Using the expression (3) and the properties of the linearized aircraft

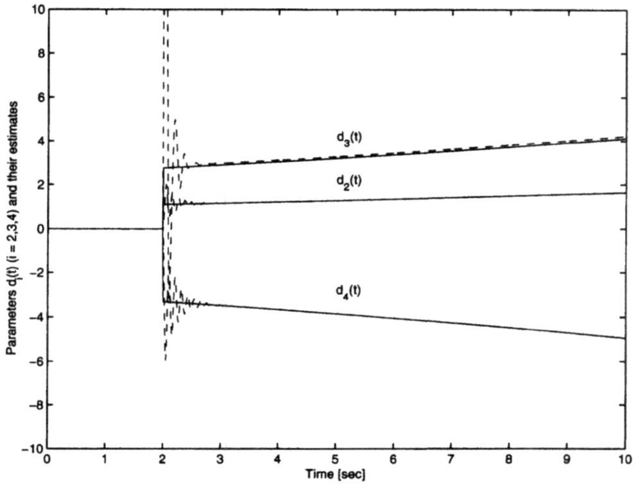

Fig. 9. *Bias parameters and their estimates (solid line: bias parameters; dashed line: estimates)*

models (see e.g. [7]), we first rewrite the plant equation (1) (for $z = 0$) in the form:

$$\dot{x}_1 = \bar{A}x, \tag{27}$$

$$\dot{x}_2 = Ax + b_1 u_{c1} + b_2 u_{c2} + \ \dots \ + b_m u_{cm}. \tag{28}$$

This is the model of the plant in the case with no failures, and can be expressed in a compact form as:

$$\dot{x}_1 = \bar{A}x, \tag{29}$$

$$\dot{x}_2 = Ax + Bu_c. \tag{30}$$

Failure Models: Using the above parametrization of u_i we next focus on the case of failures of control effectors. It is seen that each case of failure can be modeled by a different model resulting in the total of m models of the form:

$$\dot{x}_{1i} = \bar{A}x_i, \tag{31}$$

$$\dot{x}_{2i} = Ax + \bar{B}_i u_c + b_i \bar{u}_i, \quad i = 1, 2, ..., m, \tag{32}$$

where $\bar{u}_i$ can assume the following values: (i) $u_{ci}(t_{Fi})$ (Lock-in-place), or (ii) $(u_i)_{min}$ or $(u_i)_{max}$ (Hard-over). Matrices $\bar{B}_i$ from the above expression are of the form:

$$\bar{B}_1 = [\underline{0} \;\; b_2 \;\; b_3 \; \dots \; b_{m-1} \;\; b_m]$$

$$\bar{B}_2 = [b_1 \;\; \underline{0} \;\; b_3 \; \dots \;\; b_{m-1} \;\; b_m]$$

$$\vdots \qquad\qquad \vdots$$

$$\bar{B}_{m-1} = [b_1 \;\; b_2 \;\; b_3 \; ... \; \underline{0} \;\; b_m]$$
$$\bar{B}_m = [b_1 \;\; b_2 \;\; b_3 \; ... \; b_{m-1} \;\; \underline{0}],$$

where $\underline{0}$ is an m-vector with zero elements. It is seen that each failure is modeled by removing the corresponding column of matrix B, and adding the term $b_i\bar{u}_i$.

FDI Observers: The corresponding FDI observers are chosen in the following form:

$$\dot{\hat{x}}_{1o} = \bar{A}\hat{x}_o, \tag{33}$$
$$\dot{\hat{x}}_{2o} = \Lambda\hat{e}_o + Ax + Bu_c, \tag{34}$$
$$\dot{\hat{x}}_{1i} = \bar{A}\hat{x}_i, \tag{35}$$
$$\dot{\hat{x}}_{2i} = \Lambda\hat{e}_i + Ax + \bar{B}_i u_c + b_i\hat{u}_i, \quad i = 1,2,...,m, \tag{36}$$

where $\hat{e}_i = \hat{x}_i - x$, $i = 0,1,2,...,m$, $\hat{e}_i = [\hat{e}_{1i}^T \;\; \hat{e}_{2i}^T]^T$, $\hat{x}_i = [\hat{x}_{1i}^T \;\; \hat{x}_{2i}^T]^T$, and denotes the estimate of u_i. Equations (33), (34) correspond to the observer for the no-failure case, while (35), (36) correspond to the observers for the m effector failures. The matrix Λ is chosen such that the matrix $\Lambda_o = [\bar{A}^T \;\; \Lambda^T]^T$ is asymptotically stable.

Let P be a symmetric positive definite solution of $\bar{\Lambda}_o^T P + P\bar{\Lambda}_o = -Q$, where $Q = Q^T > 0$.

On-line Estimation of $\hat{u}_i$: Adaptive algorithms for adjusting $\hat{u}_i(t)$ are chosen in the form:

$$\dot{\hat{u}}_i = -\gamma_i\hat{e}_i^T P b_i, \quad i = 1,2,...,m, \tag{37}$$

where $\gamma_i > 0$ denote adaptive gains.

To increase robustness of the system, we can also adjust the estimates $\hat{u}_i$ using adaptive algorithms with projection of the form:

$$\dot{\hat{u}}_i = \mathrm{Proj}_{[(u_i)_{min},(u_i)_{max}]}\{-\gamma_i\hat{e}_i^T P b_i\}, \;\; \hat{u}_i(0) \in [(u_i)_{min}, (u_i)_{max}], \tag{38}$$
$$i = 1,2,...,m,$$

where $\mathrm{Proj}\{\cdot\}$ denotes the projection operator.

Our objective is to design control algorithms corresponding to the above FDI observers, and devise a suitable strategy for switching among the controllers to achieve the control objective in the presence of the above discussed failures. The baseline control strategy is discussed in the following section.

5.2 The Baseline Reconfigurable Controller

To derive the reconfigurable control laws for the above cases, we first rewrite the baseline control law (5) in the form:

$$u_c = W^{-1}B^T(BW^{-1}B^T)^{-1}\eta, \tag{39}$$

where

$$\eta = -Ax + A_m x_m + B_m r. \tag{40}$$

Based on the above failure models and the approach from [12], we design $m + 1$ controllers of the form:

$$u_{ci} = \Theta_i(\eta - b_i \bar{u}_i), \ i = 1, 2, ..., m, \tag{41}$$

where $\Theta_i \in \mathrm{IR}^{m \times k}$ and is chosen as [12]:

$$\Theta_i = W^{-1} \bar{B}_i^T [\bar{B}_i W^{-1} \bar{B}_i^T]^{-1}, \ i = 1, 2, ..., m. \tag{42}$$

As shown in [12,7], such controllers achieve the objective in the presence of known control effector freezing and hard-over.

The above controllers are designed for the ideal case, i.e. assuming that the type of failure (i.e. the value of $\bar{u}_i$) is known. The corresponding adaptive reconfigurable controllers are based on on-line estimation of $\bar{u}_i$, and are of the form

$$u_{ci} = \Theta_i(\eta - b_i \hat{u}_i), \ i = 1, 2, ..., m, \tag{43}$$

where $\hat{u}_i$ is the estimate of $\bar{u}_i$.

Our next objective is to devise a suitable strategy for switching among these controllers so that the control objective is achieved in the presence of the above discussed failures.

5.3 Multiple Model Adaptive Controller

In [7], a reconfigurable control strategy was designed that is based on switching among the above controllers, and that results in a stable overall system. The scheme is discussed below.

The switching scheme: Switching among the controllers is based on the following performance indices:

$$I_j(t) = c_1 \|\hat{e}_j(t)\|^2 + \frac{c_2}{t + c_3} \int_{t_o}^{t} \|\hat{e}_j(\tau)\|^2 d\tau, \ j = 0, 1, 2, ..., m, \tag{44}$$

where $\hat{e}_j = \hat{x}_j - x$, $c_i > 0$, $i = 1, 2, 3$.

The scheme is started with u_{co}, and is implemented by calculating and comparing the above indices every t_s instants, and finding their minimum. Once the minimum is found, the scheme switches to (or stays at) the corresponding controller.

In this section we will refine the above switching scheme by introducing a suitable threshold such that, in the case when one or more indices have the same value as that for the no-failure observer, the scheme chooses the latter. In addition, the state estimates generated by the FDI observers are reset to $x(t)$ every t_s instants.

Stability Analysis: The proof of stability of the overall system was given in [7]. However, it is based on an assumption that both u and $\dot{u}$ are known in advance to be bounded. We will relax this assumption here, and also prove explicitly that, in the case of a failure, the corresponding estimate converges to its true value.

Theorem 4. *In the case of failure of the ith effector, the above switching scheme assures the stability of the system (27), (28), (33)-(36), (37) and (43) and guarantees that:*
(a) $\lim_{t\to\infty}[\hat{u}_i(t) - u_i(t)] = 0$; *and*
(b) $\lim_{t\to\infty}[x(t) - x_m(t)] = 0$.

Proof: The proof of this theorem can be found in [9].
Comment: Even though it can be proved in a straightforward fashion that $\hat{u}_i(t)$ will converge to zero asymptotically, it is clear that the other estimates will not converge to their true values even if $u_c(t)$ converges to a constant vector. This is due to the fact that each observer assumes that the no-failure case is the nominal regime, while, immediately following a failure of the ith effector, the new nominal regime is the failure mode of the ith effector. Hence, the scheme cannot handle subsequent failures which can cause significant performance deterioration and even instability of the system. For these reasons, we designed a hybrid FDI-ARC scheme. This is discussed in the following section.

5.4 Hybrid FDI-ARC Scheme

The Hybrid FDI-ARC system is shown in Figure 10. The overall system is managed as an event-driven system. In particular, to increase its robustness and decrease computing time, the Decision-Making Subsystem monitors the system behavior on-line and takes appropriate actions. These include freezing of the parameter estimates associated with failures, disturbance or structural damage parameters, and resetting all other observers to the new operating regime immediately following an upset condition. The advantage of this approach is that most of the time only one fixed observer is run, while multiple observers are run only a short time immediately following failure detection and until the pre-specified settling time for the parameter estimate is reached. This substantially increases the robustness of the overall system, and decreases computing time.

We will next discuss our hybrid FDI-ARC design. Our first objective is to address the speed of convergence of the position estimate of the failed effector to its true value. Following that, we will show how to use the convergence time to reset all other observers to the new set of values. The design is carried out in several steps as discussed below.
Step 1: In this step we will simplify the FDI-ARC design from [7]. We note that the vector x_1 from (27), (28) is not directly affected by u. Hence,

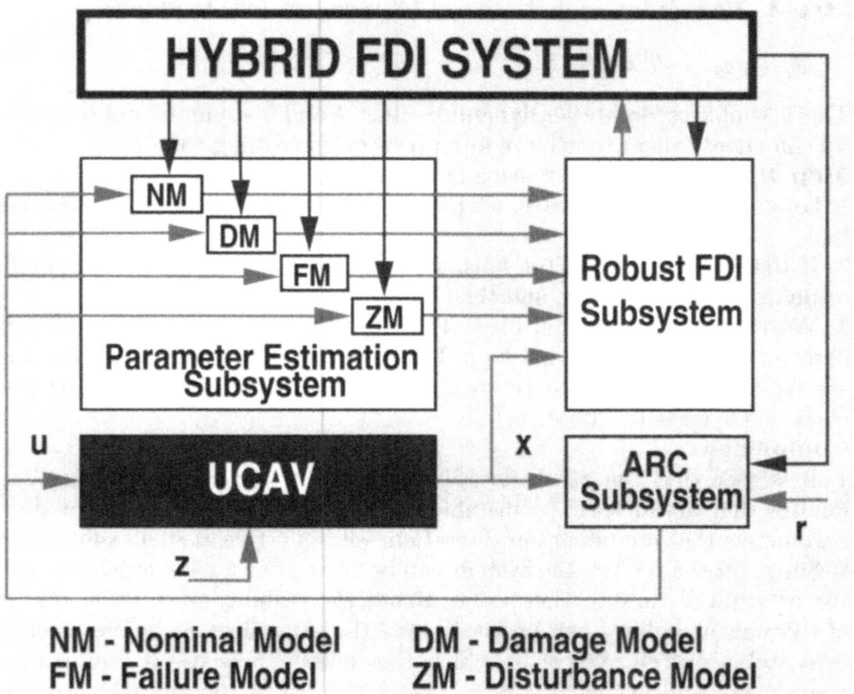

for estimating control effector failure it is sufficient to consider the relative-degree-one subsystem, i.e. the x_2 subsystem. Hence the FDI observers can be chosen as:

$$\dot{\hat{x}}_{2o} = -\lambda \hat{e}_{2o} + Ax + Bu_c, \tag{45}$$

$$\dot{\hat{x}}_{2i} = -\lambda \hat{e}_{2i} + Ax + \bar{B}_i u_c + b_i \hat{u}_i, \quad i = 1, 2, ..., m, \tag{46}$$

where $\lambda > 0$ is a scalar, and $\hat{u}_i$ is adjusted using:

$$\dot{\hat{u}}_i = -\gamma_i \hat{e}_i^T b_i, \ \ i = 1, 2, ..., m. \tag{47}$$

It can be readily verified that the above adaptive algorithm results in a stable overall system in which the control objective is achieved.
Step 2: Let the ith effector fail. We now subtract (32) from the expression (46) to obtain:

$$\dot{\hat{e}}_{2i} = -\lambda \hat{e}_{2i} + b_i \phi_i, \quad i = 1, 2, ..., m, \tag{48}$$

where $\phi_i = \hat{u}_i - \bar{u}_i, \ i = 1, 2, ..., m.$

Step 3: We now take a derivative of (47) and use (48) to obtain:

$$\ddot{\hat{\phi}}_i + \lambda\dot{\phi}_i + \gamma b_i^T b_i \phi_i = 0. \tag{49}$$

This is simple second-order dynamics. Since λ and γ are under our discretion, we can choose them to achieve any prespecified settling time τ_s.
Step 4: The supervisory scheme now works as follows:
1. For each of the i estimates, we pre-specify the corresponding settling time τ_{si}.
2. If the ith control effector fails, starting at t_o we run the corresponding estimator for τ_{si} instants, and then freeze the estimate at $t_o + \tau_{si}$.
3. We remove the nominal observer from the scheme, and reset all other observers so that the terms $b_i u_i$ that correspond to the ith control input are replaced with constant terms $b_i \hat{u}_i(t_o + \tau_{si})$. In addition, the state of all observers is reset to $x(t_o + \tau_{si})$.
Comments:
1. It is seen that the above described scheme is well suited for the case of failures that are at least τ_{si} instants apart. Since τ_{si} depends on the design parameters that are under our discretion, we can arrive at small values of τ_{si}.
2. Since, after a failure, the system will be reset after τ_{si} instants, we can set the new initial time to that value. Hence the stability analysis in the case of subsequent failures can proceed along the same lines as before, resulting in a stable overall system in which the boundedness of all the signals is guaranteed, and asymptotic convergence to zero of the tracking error and parameter estimates is guaranteed.

We will next present the simulation results for the F/A-18C/D aircraft controlled by the proposed FDI-ARC control system. This is discussed in the following section.

5.5 Simulations

The performance that can be achieved using the proposed FDI-ARC scheme is evaluated on the semi-nonlinear model of F/A-18C/D dynamics. The test vehicle is described in [11], and its response in the nominal (i.e. no-failure) case is shown in Figure 6.

Simulation 4: Uncompensated Multiple Failures. *We first simulated the aircraft with the baseline controller in the case of multiple failures. The failure scenario chosen is such that the left rudder goes hardover at t* = 1.5 *seconds, and the right LEF freezes at t* = 3.5 *seconds. The resulting response is shown in Figure 11. It is seen that the response is unacceptable and that control reconfiguration is mandatory in order to achieve the control objective.*

Simulation 5: Proposed FDI-ARC in the Presence of Multiple Failures. *We next simulated the same failure scenario, but this time with the*

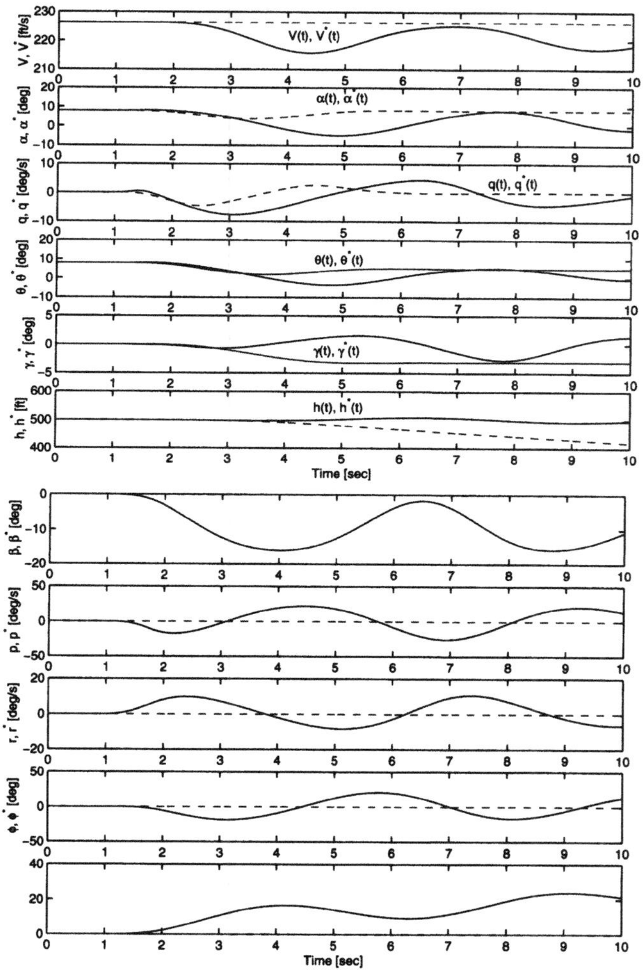

Fig. 11. *State response with the baseline controller in the case when the rudder fails at* $t = 1.5$ *and LEF fails at* $t = 3.5$ *(solid line: actual response; dashed line: desired response)*

proposed Hybrid FDI-ARC scheme. The total of seven models was built, one for the nominal (no-failure) case, and six for the following control effectors: Left and Right Leading Edge Flap, Left and Right Stabilator. and Left and Right Rudder. The following parameter values were chosen for the FDI-ARC: $\gamma = 5000, \lambda = 40, c_1 = 1, c_2 = 10,$ *and* $c_3 = 100$. *The response is shown in Figures 13-14. It is seen that the actual response is barely distinguishable from the desired one, and that the control inputs are reconfigured to compensate for the failure. This demonstrates the feasibility of the proposed approach. In the*

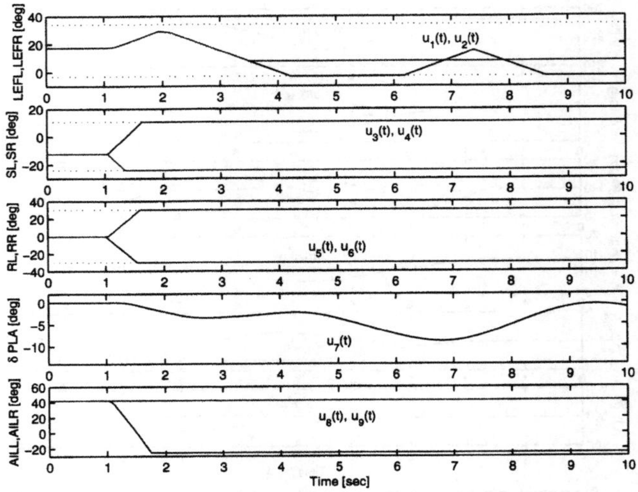

Fig. 12. *Input Response with the baseline controller in the case when the rudder fails at* $t = 1.5$ *and LEF fails at* $t = 3.5$ *(dashed lines denote position limits)*

right part of the Figure 14 we plotted the estimate of the failed control input, and the switching sequence that shows as to how the system detects the failure and resets to the new nominal regime established immediately following each failure.

5.6 Conclusions

In this section we present a new Failure Detection and Identification (FDI) and Adaptive Reconfigurable Control (ARC) scheme for achieving the desired flight performance in the presence of multiple control effector failures that occur short time apart. The failures include lock-in-place and hard-over. The problem of stable control reconfiguration in the presence of such failures is solved using an scheme employing multiple adaptive FDI observers and controllers and a suitably chosen decision-making mechanism. The scheme is shown to guarantee the detection and identification of a failed effector. Due to a convenient structure of the overall system, the observers can be modified on-line to reflect the new operating regime following the failure. This enables fast FDI and effective ARC in the case of subsequent failures. The stability of the overall FDI-ARC system is demonstrated using the Lyapunov method.

6 Structural Damage FDIR

In this section we will consider a linearized model of a Tailless Advanced Fighter Aircraft (TAFA) in the presence of wing damage. This aircraft is

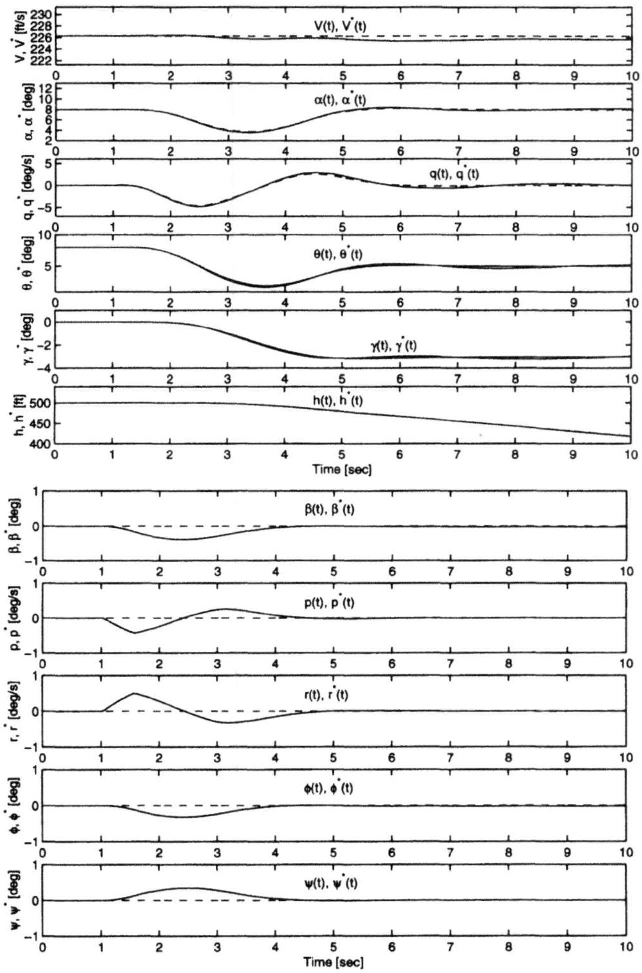

Fig. 13. *State response with the proposed adaptive reconfigurable controller (solid line: actual response; dashed line: desired response)*

a variant of X-36, and is shown in Figure 15. Its main feature is that is is equipped by a whole suite of innovative control effectors resulting in substantially increased control input redundancy and reconfiguration capabilities [13]. Such a redundancy is effectively used here to compensate for severe damage of the outer portion of the wing.

The generalized aircraft model is of the form:

$$\dot{x} = A_D(t)x + B_D(t)u, \tag{50}$$

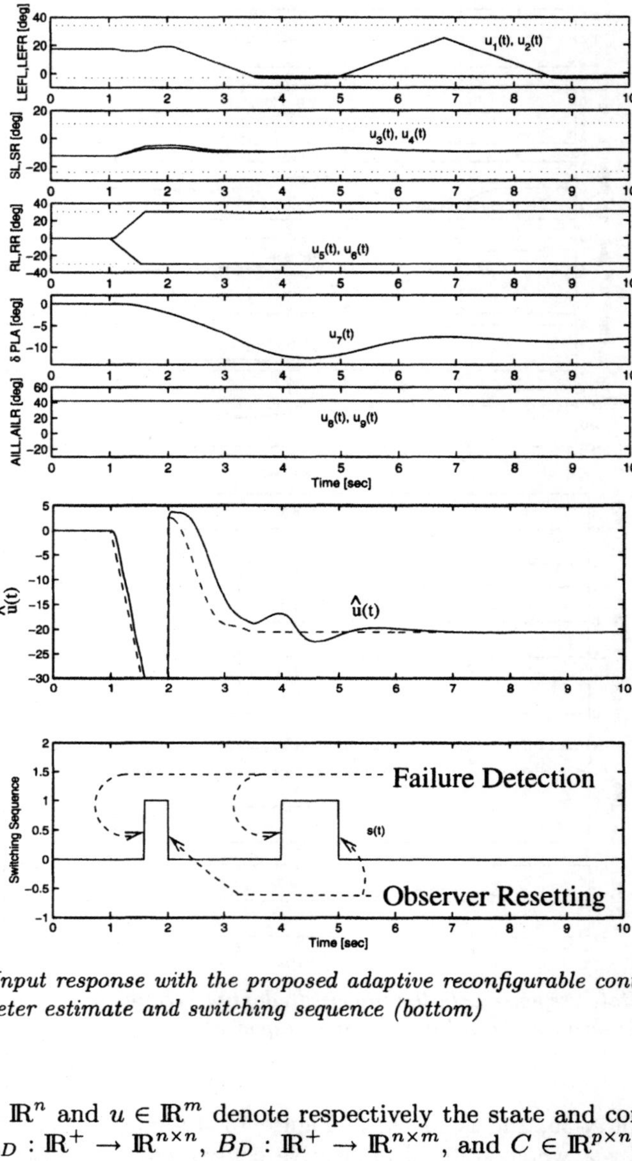

Fig. 14. *Input response with the proposed adaptive reconfigurable controller (top) and parameter estimate and switching sequence (bottom)*

where $x \in \mathbb{R}^n$ and $u \in \mathbb{R}^m$ denote respectively the state and control input vectors, $A_D : \mathbb{R}^+ \to \mathbb{R}^{n\times n}$, $B_D : \mathbb{R}^+ \to \mathbb{R}^{n\times m}$, and $C \in \mathbb{R}^{p\times n}$. The time varying nature of matrices A_D and B_D is due to the wing damage which affects the dynamics of the aircraft in an abrupt fashion. In [13], the following form of these matrices was proposed:

$$A_D(\rho(t)) = A_{Do} + A_{D1}\rho(t) + A_{D2}\rho^2(t) + A_{D3}\rho^3(t) + A_{D4}\rho^4(t) + A_{D5}\rho^5(t), \quad (51)$$

$$B_D(\rho(t)) = B_{Do} + B_{D1}\rho(t) + B_{D2}\rho^2(t) + B_{D3}\rho^3(t) + B_{D4}\rho^4(t) + B_{D5}\rho^5(t), \quad (52)$$

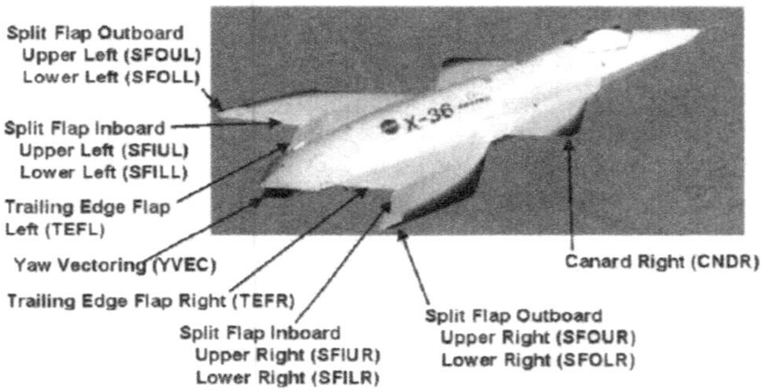

Fig. 15. *The Tailless Advanced Fighter Aircraft (TAFA)*

where $\rho \in [0, 1]$ denotes the damage parameter. The above model sufficiently accurately covers all cases from $\rho = 0$ (no-damage case) to $\rho = 1$ (100% wing damage) [13].

In this section our objective is to design a suitable control strategy for the above aircraft model such that the desired performance is achieved despite the presence of severe wing damage. In particular, our strategy will include guaranteed robustness along with on-line reconfiguration in the case when $\rho(t)$ abruptly switches within the set $[0, 1]$.

We will first make the following assumption:

Assumption 2: There exists a nonsingular transformation matrix T that transforms the model (50) to the form:

$$\dot{x}_1 = \bar{A}x_2, \quad (53)$$
$$\dot{x}_2 = A(t)x + B(t)u, \quad (54)$$

where x_1 is an $(n-p)$-vector, x_2 is a p-vector, $\bar{A}$ is a $((n-p) \times p)$ matrix independent of ρ, A and B are respectively $p \times n$, and $p \times m$ matrices, and

$$A(\rho(t)) = A_o + A_1\rho(t) + A_2\rho^2(t) + A_3\rho^3(t) + A_4\rho^4(t) + A_5\rho^5(t), \quad (55)$$
$$B(\rho(t)) = B_o + B_1\rho(t) + B_2\rho^2(t) + B_3\rho^3(t) + B_4\rho^4(t) + B_5\rho^5(t). \quad (56)$$

We will consider the case when the desired behavior of the plant is specified by a reference model of the form:

$$\dot{x}_{m1} = \bar{A}x_{m2}, \quad (57)$$
$$\dot{x}_{m2} = A_m x_m + B_m r, \quad (58)$$

where $x_{m1} \in \mathbb{R}^(n-p)$, $x_{m2} \in \mathbb{R}^p$, $[\bar{A}^T \; A_m^T]^T$ is an asymptotically stable $(n \times n)$ matrix, B_m is a $(n \times p)$ matrix, and r denotes a p-vector of bounded piecewise continuous reference inputs.
Considering the plant and reference model representations we have that $x = [x_1^T \; x_2^T]^T$, and $x_m = [x_{m1}^T \; x_{m2}^T]^T$. The control objective is now formally stated as follows:
Control Objective: Design a control law $u(t)$ for the plant (53), (54) such that all signals in the system are bounded, and $\lim_{t\to\infty}[x(t) - x_m(t)] = 0$, in the presence of variations of $\rho(t)$ within $[0, 1]$ that satisfy assumption 1(v).

Hence the requirement is, even in the case of 100% wing damage, to achieve the same level of performance as in the no-damage case.

6.1 Baseline Control Strategy

Since, by virtue of assumption 1, the plant is invertible, we suggest the control law based on dynamic inversion for the overactuated case, which is also referred to as the pseudo-inverse control [17]. We will refer to this control law as the Inverse Dynamics Control Law (IDCL).

Let $\eta(x, \alpha, t) = \{-A(\alpha)x + \Lambda(x - x_m) + A_m x_m + B_m r\}$. The control law is now of the form:

$$u = W^{-1}B^T(\alpha)[B(\alpha))W^{-1}B^T(\alpha)]^{-1}\eta(x, \alpha, t), \tag{59}$$

where W is a diagonal $(m \times m)$ matrix with strictly positive elements, and Λ is chosen such that the matrix

$$\Lambda_o = \begin{bmatrix} \bar{A} \\ \Lambda \end{bmatrix} \tag{60}$$

is asymptotically stable. The IDCL law also minimizes the total control effort, as discussed in [17] (see the Appendix).

We note that the above IDCL is augmented with the output error feedback term $\Lambda(x - x_m)$.

The above control law can be readily shown to achieve the objective for a known operating regime of the plant. Let $e = x - x_m$ denote the output error vector. From (53), (54), (4), and (59) we obtain that $\dot{e} = \Lambda_o e$. This system is exponentially stable since Λ_o is asymptotically stable. In addition, Λ_o can be chosen to place the poles of the closed-loop system arbitrarily far into the left half plane.

The above control laws assume complete prior knowledge of matrices A and B. Since this assumption is not realistic in the case when $\rho > 0$, we will further consider a MMST-based control design for the model (53), (54).

6.2 Adaptive Control Design

In this section we will discuss the design of an on-line scheme for estimation of ρ, as well as the corresponding reconfigurable controller, and study the stability of the overall system.

We recall that the TAFA model is of the form (53), (54), where A and B are given by (55) and (56). Since the parameter ρ is unknown (i.e. the extent of damage and its time instant of occurrence is not known), we next build an observer for estimating ρ on-line and design a corresponding controller, as shown below.

Observer: The observer is chosen in the form:

$$\dot{\hat{x}}_1 = \bar{A}\hat{x}_2, \tag{61}$$

$$\dot{\hat{x}}_2 = \Lambda\hat{e} + \hat{A}x + \hat{B}u, \tag{62}$$

where $\hat{e} = \hat{x} - x$, $\hat{x} = [\hat{x}_1 \quad \hat{x}_2]^T$, and

$$\hat{A} = A_o + A_1\hat{\rho} + A_2\hat{\rho}^2 + A_3\hat{\rho}^3 + A_4\hat{\rho}^4 + A_5\hat{\rho}^5, \tag{63}$$

$$\hat{B} = B_o + B_1\hat{\rho} + B_2\hat{\rho}^2 + B_3\hat{\rho}^3 + B_4\hat{\rho}^4 + B_5\hat{\rho}^5, \tag{64}$$

where $\hat{\rho}$ denotes the estimate of ρ.

Controller: Let $\hat{\eta} = C_j\{-\hat{A}x + \Lambda(x - x_m) + A_m x_m + B_m r\}$. The baseline control law is now modified as follows:

$$u = W^{-1}(C_j\hat{B})^T[C_j\hat{B}W^{-1}((C_j\hat{B})^T]^{-1}\hat{\eta}. \tag{65}$$

On-line estimation of ρ: Since it was assumed that $\rho \in [0,1]$, we will also adjust its estimate within the interval $[0,1]$ using the adaptive algorithms with projection (see the Appendix). This will also assure that the properties (iii) and (iv) from the assumption 2 are retained even when the true value of ρ is substituted by its estimate in the control law.

Let $\mathcal{S}$ denote the system consisting of the plant (53), (54) and the controller (65). Let also P denote a symmetric positive definite solution of the Lyapunov matrix equation $\Lambda_o^T P + P\Lambda_o = -Q$, where $Q = Q^T > 0$ and Λ_o is given by (60). We next consider the following theorem:

Theorem 5. *If the estimate $\hat{\rho}$ is adjusted using the following adaptive law:*

$$\dot{\hat{\rho}} = Proj_{[0,1]}\{-\gamma\hat{e}^T P\Omega(x,u,\hat{\rho})\}, \tag{66}$$

where

$$\Omega(x,u,\hat{\rho}) = \sum_{i=1}^{5} i\cdot[A_i x + B_i u]\hat{\rho}^{i-1}, \tag{67}$$

and $\gamma > 0$, then all the signals in the system $\mathcal{S}$ are bounded and $\lim_{t\to\infty}[x(t) - x_m(t)] = 0$ in the presence of arbitrary variations of $\rho(t)$ within $[0,1]$.

Proof: The prof of this theorem can be found in [8].

Comment:

1. Due to the properties of adaptive algorithms with projection [21], we can readily demonstrate that the system will be robust in the presence of large bounded external disturbances, noise and some classes of unmodeled dynamics.

2. Since an error model linearized with respect to ϕ was used to design adaptive algorithms, the stability property of the overall system is local in nature, i.e. the approximation is valid for sufficiently small ϕ. Global stability can be demonstrated using an approach where the adaptive controller is combined with a variable-structure controller using multiple models. In the problem under consideration, numerous simulations revealed that the values $\phi \in [0, 1]$ are sufficiently small to guarantee the validity of the approximation and assure excellent overall performance. Some representative simulations are included in the following section.

6.3 Simulations

Since the TAFA dynamics is open-loop unstable, any percentage of damage results in unacceptable response. To test our adaptive reconfigurable algorithm, we used Boeing's high-fidelity 6DOF TAFA simulation that has a damage simulation capability. A detailed description of the corresponding implementation issues is given in [6] and [5]. We will include here some representative simulation results. As the simulation test case we have chosen the flight regime of Mach 0.9 at the sea-level. The test maneuver is altitude capture under wing damage.

Simulation 6: Nominal response. *Before we test the suitability of our FDI-ARC scheme for handling wing damage, it is important to demonstrate that the scheme does not give false alarm and/or interfere with the nominal controller in the case with no damage. Figure 16 shows the state response of the aircraft under the no-damage condition. The bottom right figure shows the parameter estimate* $\hat{\rho}$. *As it can be seen, the scheme identifies no wing damage and* $\hat{\rho}$ *remains zero during the simulation. The nominal controller is used at all time and gives excellent tracking response.*

Simulation 7: Response in the case of 60% outer wing damage. *Figure 17 shows the aircraft state response and control effector positions in the case of a 60% wing damage that occurs at* $t = 1.5$ *sec. The parameter estimate* $\hat{\rho}$ *converges to 0.645 after 10 sec. The control switched to* u_{f2} *as the failure was detected. This results in stable tracking performance of the aircraft. It should be noted that if no configuration is attempted, the closed-loop system becomes unstable and the system states rapidly diverge.*

Simulation 8: Response in the case of 100% outer wing damage. *Figure 18 shows the aircraft state response and control effector positions in the case of 100% wing damage that occurs at* $t = 1.5$ *sec. The parameter estimate* $\hat{\rho}$ *converges to 0.916 after 10 sec. The control switched to* u_{f3} *as the failure is detected. Again, stable tracking performance of the aircraft is achieved. The FDI-ARC scheme was tested extensively for different levels of wing damage under the condition described above. In all cases, the parameter estimate* $\hat{\rho}$ *converges within* ± 0.1 *of the actual* ρ, *and the control is always reconfigured in a manner that achieves stability and good tracking performance.*

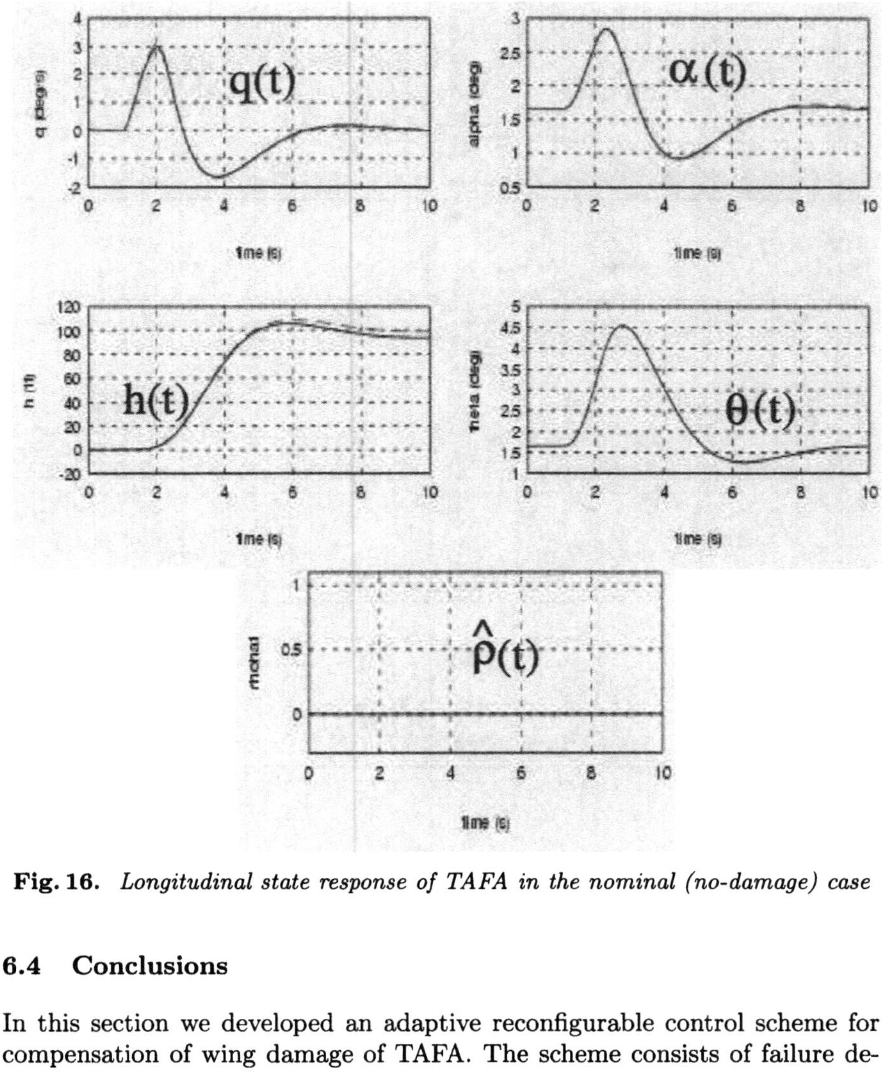

Fig. 16. *Longitudinal state response of TAFA in the nominal (no-damage) case*

6.4 Conclusions

In this section we developed an adaptive reconfigurable control scheme for compensation of wing damage of TAFA. The scheme consists of failure detection through on-line estimation of the percentage-of-damage parameter, and the corresponding adaptive reconfigurable control algorithm. The overall scheme results in fast and accurate detection and identification of wing damage and corresponding control reconfiguration, and was demonstrated analytically to be stable in the sense that all signals are bounded and the output error converges to zero asymptotically even in the case of 100% wing damage. The properties of the proposed adaptive controller are evaluated through numerical simulations on a high-fidelity TAFA simulation, and it is shown that the proposed approach results in excellent overall performance despite severe structural damage.

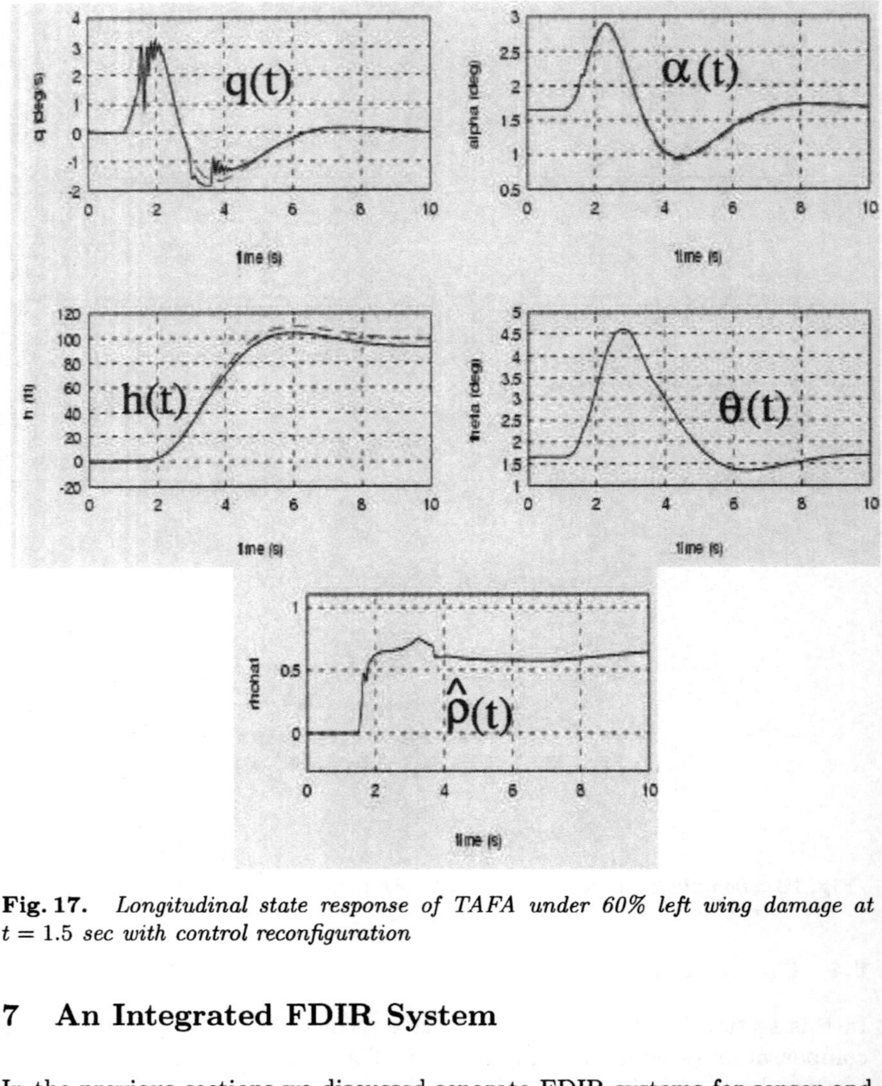

Fig. 17. *Longitudinal state response of TAFA under 60% left wing damage at* $t = 1.5$ *sec with control reconfiguration*

7 An Integrated FDIR System

In the previous sections we discussed separate FDIR systems for sensor and control effector failures, and structural damage. While each individual system achieves its own objective, the overall objective is to compensate for a wide array of upsets. These systems, therefore, need to be implemented simultaneously, which is a highly complex problem due to strong inter-system couplings. When such a multi-objective task is to be achieved under state, output, and control input constraints, the problem at hand becomes truly formidable. The related control problem is that of system integration that needs to achieved without violating the constraint and desired dynamics specifications.

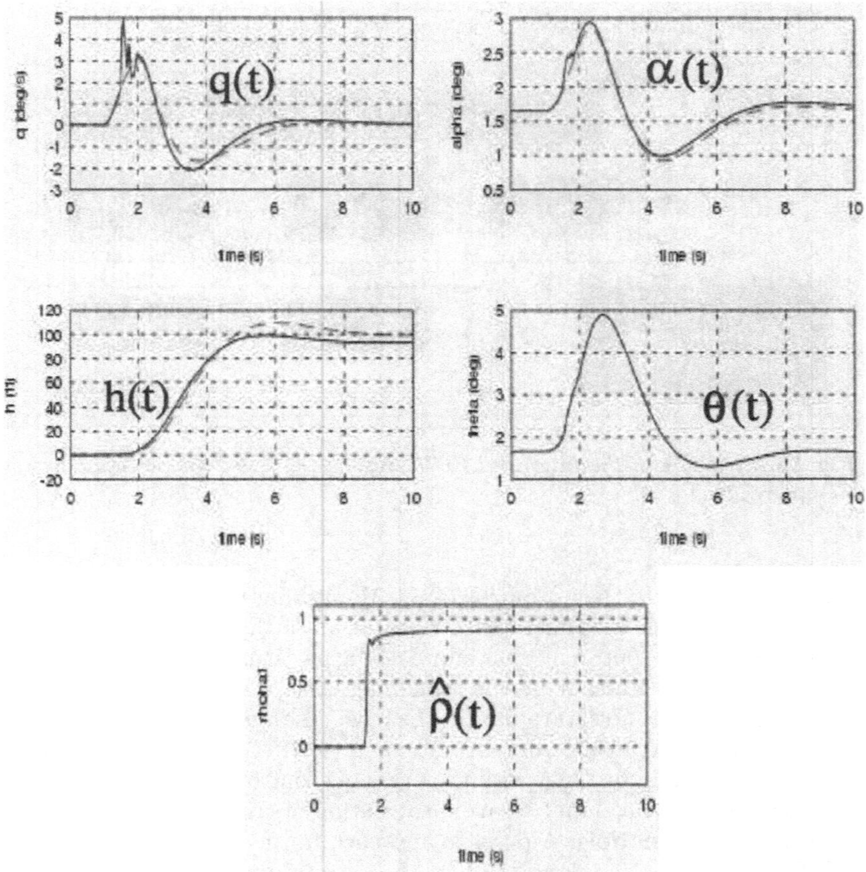

Fig. 18. *Longitudinal state response of TAFA under 100% left wing damage at* $t = 1.5$ *sec with control reconfiguration*

A conceptual solution for such an Integrated Failure Detection, Identification and Reconfiguration (FDIR) system is shown in Figure 1. The system needs to have a capability of compensating for *multiple simultaneous sensor and control effector failures and structural damage.* The FDIR algorithms need to distinguish between different failures and rapidly and accurately identify failure-related parameters.

Control Input Monitoring and Allocation (CIMAS) system: One of the key issues in reconfigurable control design is that of control input allocation before and after the reconfiguration. The Control Allocation Algorithms (CAA) pre and post failure can be very different, and finding the adequate CAA for different combinations of failures is a formidable task. For this rea-

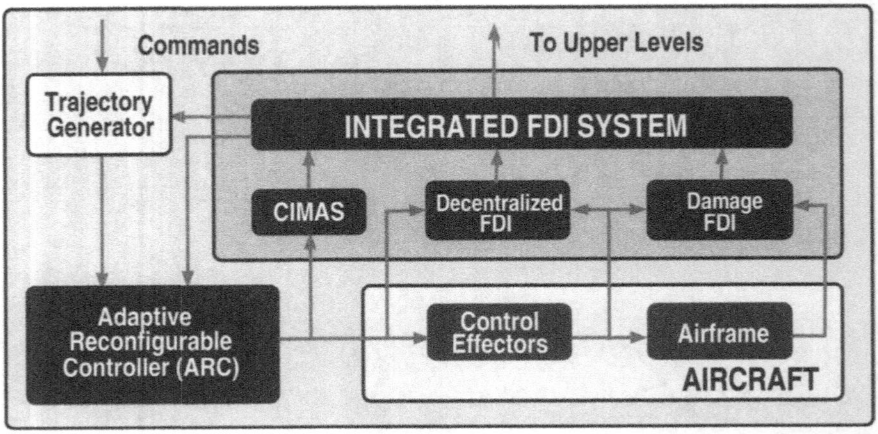

Fig. 19. *Structure of the Integrated FDIR System (*©1999-2002 Scientific Systems Company, Inc.*)*

son, we developed a new Control Input Monitoring and Allocation System (CIMAS). The role of CIMAS is to monitor all of the control inputs, detect when a particular control input is approaching its saturation bound, and carry out two tasks: (i) Re-allocate the remaining control inputs automatically so that either there is no saturation (in the ideal case), or, if there is saturation, then only non-critical effectors saturate; and (ii) If there is no solution to the problem of finding the CAA with the available control authority, the CIMAS changes the reference input to avoid saturation of critical effectors.

Failure/upset models: A possible approach to the design of an Integrated FDIR system uses the description of the system dynamics of the form (1), (2). The modeling starts with sensor failures:

$$\begin{aligned} y &= Cx + d \\ \dot{y} &= C\dot{x} = CAx + CBu \\ &= CAC^{-1}(y - d) + CBu \end{aligned}$$

We next replace the term Bu in the case of the control effector failures with $BK(I - \Sigma)u + B\Sigma\bar{u}$, where K is the control effector effectiveness matrix described earlier, $\Sigma = \text{diag}[\sigma_1\ \sigma_2\ ...\ \sigma_m]$, $\bar{u}$ denotes an (unknown) vector whose elements are positions at which different effectors have locked, and

$$\sigma_i(t) = \begin{cases} 1, \text{ if } t \geq t_{Fi}, \\ 0 \text{ elsewhere} \end{cases}$$

and t_{Fi} denotes the time of failure of the ith effector. Hence the combined sensor-control effector failure model is of the form:

$$\dot{y} = CAC^{-1}y - CAd + C[BK(I - \Sigma)u + B\Sigma\bar{u}].$$

We further assume that matrices A and B change due to structural damage in a similar way as described in the section on damage FDIR. However, the model (55), (56) describes only structural damage of the outer part of the left wing of TAFA. In order to take into account damage of different surfaces, we need to introduce a vector ρ such that $\rho = [\rho_1 \; \rho_2 \; ... \; \rho_N]$, where N is the number of surfaces that can be damaged. The model is now of the form similar to (55), (56) except that different order polynomials may have to be used to model damage of different surfaces. Hence the overall model is of the form:

$$\dot{y} = CA(\rho)C^{-1}y - CA(\rho)d + C[B(\rho)K(I - \Sigma)u + B(\rho)\Sigma\bar{u}].$$

We need to emphasize that the above model does not take into account failures a number of important subsystems including the flight computer, communication channels, and power system. Hence the proposed FDIR system needs to be combined with an on-line system-wide health monitoring and status determination system that can rapidly and accurately assess the overall health of the vehicle, and take appropriate actions in order to minimize the effect of different failures and upsets on the closed-loop system.

8 Conclusions and Future Work

In the paper we describe issues arising in Failure Detection, Identification and Reconfiguration (FDIR) in flight control. The fault accommodation problem is stated and discussed, followed by the derivation of models of sensor and control effector failures and structural damage. Corresponding FDIR schemes that minimize the effect of different failures and structural damage are presented next, and their properties are illustrated through numerical simulations of advanced combat aircraft. At the end, an Integrated FDIR scheme is discussed consisting of on-line health monitoring and status determination system combined with the proposed FDIR algorithms.

The design of such a comprehensive health monitoring and FDIR system will be the focus of research in this area. The techniques that will arise from this research are expected to eventually lead to the development and implementation of a fully autonomous Intelligent Vehicle Management System (IVMS).

References

1. Ahmed-Zaid F., Ioannou P., Gousman K., Rooney R., (1991) Accommodation of Failures in the F-16 Aircraft Using Adaptive Control. IEEE Control Systems Magazine 11(1):73–78
2. Bodson M. and Groszkiewicz J. (1997) Multivariable Adaptive Algorithms for Reconfigurable Flight Control. IEEE Transactions on Control Systems Technology 5:217–229

3. Boeing Phantom Works (1998) Reconfigurable Systems for Tailless Fighter Aircraft - RESTORE (First Draft). Contract No. F33615-96-C-3612, Scientific and Technical Reports, System Design Report, CDRL Sequence No. A007, St. Louis, Missouri
4. Bošković J.D. (1998) Adaptive Control of a Class of Nonlinearly-Parametrized Plants. IEEE Transaction on Automatic Control 43:930-934
5. Bošković J.D., Li S.-M., Mehra R.K. (2000) Reconfigurable Flight Control Design Using Multiple Switching Controllers and On-line Estimation of Damage-Related Parameters. Proc. of 2000 Conference on Control Applications, Anchorage, Alaska
6. Bošković J.D., Li S.-M., Mehra R.K. (2000) Evaluation of the Properties of a Multiple-Model Reconfigurable Flight Controller on a 6 DOF Simulation. Proc. of 2000 AIAA Guidance, Navigation and Control Conference, Denver, CO
7. Bošković J.D., Mehra R.K. (1999) Stable Multiple Model Adaptive Flight Control for Accommodation of a Large Class of Control Effector Failures. Proc. of 1999 American Control Conference, San Diego, CA
8. Bošković J.D., Mehra R.K. (2000) Intelligent Adaptive Control of a Tailless Advanced Fighter Aircraft under Wing Damage. AIAA Journal of Guidance, Control & Dynamics 23:876–884
9. Bošković J.D., Mehra R.K. (2001) A Multiple Model Adaptive Flight Control Scheme for Accommodation of Actuator Failures. To appear in: AIAA Journal of Guidance, Control & Dynamics
10. Bošković J.D., Mehra R.K. (2002) Stable Adaptive Multiple Model-based Control Design for Accommodation of Sensor Failures. 2002 American Control Conference
11. Bošković J,D., Li S.-M., Mehra R.K. (2001) A Hybrid Fault-Tolerant Scheme for Flight Control Applications. Proc. of 2001 AIAA Guidance, Navigation and Control Conference, Montreal, Canada
12. Bošković J.D., Yu S.-H., Mehra R.K. (1998) A Stable Scheme for Automatic Control Reconfiguration in the Presence of Actuator Failures. Proc. of 1998 American Control Conference, Philadelphia, PA, 2455–2459
13. Brinker J., Wise K. (1998) Reconfigurable Flight Control of a Tailless Advanced Fighter Aircraft. Proc. of 1998 AIAA Guidance, Navigation and Control Conference, Boston, MA, 1:75-87
14. Calise A., Lee S., Sharma M. (1998) Direct Adaptive Reconfigurable Control of a Tailless Fighter Aircraft. Proc. of 1998 AIAA Guidance, Navigation and Control Conference, Boston, MA, 1:88-97
15. Chandler P., Pachter M., Mears M. (1995) System Identification for Adaptive and Reconfigurable Control. Journal of Guidance, Control & Dynamics 18:516-524
16. Enns D. (2000) Control Allocation Approaches. Proc. of 1998 AIAA Guidance, Navigation and Control Conference, Boston, MA
17. Honeywell Technology Center (1996) Multivariable Control Design Guidelines. Report for the Program "Design Guidelines for Application of Multivariable Control Theory to Aircraft Laws", Minneapolis, MN
18. LaSalle J.P. (1960) Some Extensions of Lyapunov's Second Method. IRE Transactions on Circuit Theory 7
19. McRuer D., Ashkenas I., Graham D. (1973) Aircraft Dynamics and Automatic Control. Princeton University Press, Princeton, NJ

20. Narendra K.S., Balakrishnan J. (1997) Adaptive Control Using Multiple Models. IEEE Transactions on Automatic Control 42:171–187
21. Narendra K.S., Annaswamy A.M. (1988) Stable Adaptive Systems. Prentice Hall Inc., Englewood Cliffs, New Jersey
22. Page A., Steinberg M. (2000) A Closed-Loop Comparison of Control Allocation Methods. Proc. of 2000 AIAA Guidance, Navigation and Control Conference, Denver, CO
23. RESTORE Final Briefing (1999), WPAFB, OH
24. Wise K., Sedwick J. (1998) Stability Analysis of Reconfigurable and Gain Scheduled Flight Control System Using LMIs. Proc. of 1998 AIAA Guidance, Navigation and Control Conference, Boston, MA, 1:118-126

Nonlinear Fault Detection for Hydraulic Systems

Martin L. Leuschen[1], Ian D. Walker[2], and Joseph R. Cavallaro[1]

Abstract. One of the most important areas in the robotics industry is the development of robots capable of working in hazardous environments. As humans cannot safely or cheaply work in these environments, providing a high level of robotic functionality is important. Our work in this area focuses on a fault detection method known as analytical redundancy, or AR. In this paper we discuss the application to a hydraulic servovalve system of our novel rigorous nonlinear AR technique. AR is a model-based state-space technique that is theoretically guaranteed to derive the maximum number of independent tests of the consistency of sensor data with the system model and past control inputs. Conventional linear AR is only valid for linear sampled data systems. However, our new nonlinear AR (NLAR) technique maintains traditional linear AR's mathematical guarantee to generate the maximum possible number of independent tests in the nonlinear domain. Thus NLAR allows us to gain the benefits of AR testing for nonlinear systems with both continuous and sampled data.

1 Introduction

The usefulness of robots in hazardous situations is highly dependent on their reliability [8–10,30]. Chemicals and radiation can damage robotic components, and many environments can be made more hazardous by actions of a malfunctioning robot. As humans usually cannot enter hazardous environments to repair or remove a failed robot, such failures can be very dangerous and costly. Thus, our team has investigated reliability issues for robots extensively [17–19,23]. A fault detection method known as analytical redundancy [4,16], or AR, is the focus of this particular paper. AR is a model-based technique that derives the maximum number of independent tests from the state-space control model of the system. These AR tests monitor the consistency of sensor data with the linearized system model and past control inputs. The tests determine whether the system is performing nominally, or is deviating from the desired plan and presumably under fault conditions. Our group has used the linear version of the AR technique successfully on electrical robotic systems in the past [30], and has also applied our nonlinear version to nonlinear hydraulic systems such as the Rosie robot discussed below.

In previous papers [19,23], we have discussed the derivation through linear AR of a suite of model based tests for the default sensor package for hydraulic wheel actuators, and introduced an approximate technique for using AR efficiently in nonlinear systems. Some of these tests are comparison of the actual

F. Caccavale, L. Villani (Eds.): Fault Diagnosis and Fault Tolerance for Mechatronic Systems, STAR 1, pp. 169-191, 2003.

system response to control inputs to the predicted response indicated by the model. These show that AR clearly monitors known behaviors of the system. The other tests uncovered by the linear AR analysis reflect higher order state interdependencies, as discussed later and in [27]. These tests give us important additional information about the system that we might not have normally examined.

These previous tests are all based at least partially on linear models of the system. However, the hydraulic valve and motor system behavior is highly nonlinear in nature, which leads to a degradation in the performance of the AR method. Linear AR can miss or improperly detect faults in this situation, so nonlinear AR techniques are highly desirable. We now show our results for rigorously extending the linear AR tests into the nonlinear realm.

2 Nonlinear Analytical Redundancy

Let us begin by defining a nonlinear state-space system with states, inputs, and outputs.

$$\begin{aligned} \dot{\underline{x}}(t) &= \underline{f}(\underline{x}(t)) + \sum_i \underline{g}_i(\underline{x}(t)) \cdot \underline{u}(t) \\ \underline{y}(t) &= \underline{h}(\underline{x}(t)) \end{aligned} \tag{1}$$

The corresponding linear system model is:

$$\begin{aligned} \dot{\underline{x}}(t) &= A\underline{x}(t) + B\underline{u}(t) \\ \underline{y}(t) &= C\underline{x}(t) \end{aligned} \tag{2}$$

NLAR uses the left null-matrix of the observability:

$$[\Omega\,(\underline{f},\underline{g})]\;[\mathcal{O}_{\triangle}(\underline{f},\underline{g})] = \underline{0} \tag{3}$$

where the model-based observability $\mathcal{O}_{\triangle}(\underline{f},\underline{g})$ is calculated using standard methods using the model equations for linear systems, and our modified 'triangular' method [16,21] for nonlinear systems. The $\triangle$ refers to this method, which reduces to the standard method in the linear case) The linear observation space is the rows of the matrix $\left[C,\, CA,\, CA^2 \ldots\right]$. For nonlinear systems, the Lie Derivative operation $L_{\underline{f}} h_i$ is used to combine the vector functions $\underline{f}, \underline{g}_i$ with elements of $\underline{h}$, and the observation space is generated from appropriate combinations of $L_{\underline{k}} h_i$, $k \in \left\{\underline{f}, \underline{g}_1, \underline{g}_2, \ldots\right\}$. For more details, see [15,16,21,22].

However, if the system is observable it is also possible to express the observability in terms of sensor readings $\underline{y}$ and control inputs $\underline{u}$ [15,16,22] in addition to the state and model. This observability will be referred to as the *'dynamically derived observability,'* or $\mathcal{O}_{\triangle DD}(\underline{y},\underline{u})$. The important aspect of this formulation is the explicit dependence of every element of it on the input-output behavior of the system as it is functioning at the time the observability is calculated. By taking the product of $\mathcal{O}_{\triangle DD}(\underline{y}, \underline{u})$ with the left null-matrix from equation 3 we can generate the suite of NLAR residual tests $\underline{R}$:

$$[\Omega(\underline{f}, \underline{g})]\,[\mathcal{O}_{\triangle DD}(\underline{y}, \underline{u})] = \underline{R} = [0] \tag{4}$$

As $\Omega(f, g)$ is derived from the system equations, while the input-output observability matrix depends on the recent sensor readings and inputs, if the system is observable the result of the above matrix multiplication can be expressed as a set of equations dependent on the known quantities of $\underline{y}$ and $\underline{u}$. If the system is behaving in accordance with the nominal model $\mathcal{O}_{\triangle DD}(\underline{y}, \underline{u})$ will be as similar to $\mathcal{O}_{\triangle}(\underline{f}, \underline{g})$ as measurement error and noise allow, and equation 4 will generate near-zero values [15,16,22]. However, if the system model has become inaccurate due to a fault changing the characteristics of the system, nonzero values will be generated, allowing the fault to be detected.

In fact, since the observability space by definition spans all that can be observed about the system using the current model, it can be shown that NLAR is guaranteed to react to any observable discrepancy [15,16,22]. The basis vectors of the observability space generated from system data (model $\mathcal{O}_{\triangle DD}(\underline{y}, \underline{u})$) span a space dependent on the current behavior of the system that spans the observability space if the model is correct. The null-space of the model derived observability space $\Omega(\underline{f}, \underline{g})$ spans the space of information you shouldn't be able to see if the system is performing according to the model. Projecting the basis onto the null space instantly reduces a complex stream of sensor and input data into residual signals that show all the deviations from the expected model, and only those deviations.

NLAR test results can only be zero if the system equations are modeling the system behavior correctly. Any discrepancies, such as those that result from sensor noise, manifest as bias or noise in the NLAR output. However, given a good system model, most faults will cause deviations between the system and the model much greater than the difference caused by modeling inaccuracies. Faults will appear as large nonzero NLAR signals, thus allowing NLAR to be used as an effective tool for fault detection [15,16,20,22].

The standard linear AR is shown to be a special case of NLAR in [16]. The linear technique described in [4] requires a linear system model like that described in equation 2. This can cause significant extra bias and noise in the linear AR tests if the system has nonlinear characteristics. NLAR can of course solve this problem.

2.1 The NLAR Algorithm

Although the full derivation of the NLAR method described above is too extensive to reproduce here, the following algorithm [16,21,22] summarizes the necessary steps in deriving a suite of NLAR tests:

i. Determine the triangular nonlinear observability $\mathcal{O}_{\triangle}$ and its left null Ω.

ii. Determine the nonlinear dynamically derived observability $\mathcal{O}_{\triangle DD}$.
iii. Find the rank, $r_j(nonlin)$, of each observability sub-matrix in the observability "matrix" in $\nabla\mathcal{O}_{\triangle DD}$. Keep $r_j(nonlin)+1$ rows in each sub-vector.
iv. Apply the NLAR equation to find the test residuals $\underline{R} : \Omega\mathcal{O}_{\triangle DD} = \underline{R}$.
v. The number of independent NLAR tests, N_{NLAR}, can be determined by the equation $N_{NLAR} = \sum_{j=1}^{m} (r_j(nonlin)) + (m - n)$. Use this equation to determine how many independent tests exist. Delete the redundant tests.

Full details of this method are discussed in [16,21,22].

2.2 Other Related Work

Considerable work has been done using the concept of nonlinear observers in place of AR [3,13]. In fact, it has been shown that AR and observer based methods are equivalent in the linear case [24]. However, the proof in [24] is only applicable to linear systems, so the nonlinear observers are not necessarily the same as nonlinear AR. Additionally, the nonlinear observer based method [13] lacks the span guarantees of NLAR.

Zhirabok and Preobragenskaya have presented work with nonlinear AR test residuals based on observer theory [33]. Nonlinear test residuals are generated by following an algorithm for restating the model equations in terms of inputs u_i and outputs y_i. Unfortunately this method does not use the observability to maintain the guarantees that make AR so desirable.

Wünnenberg and Frank have investigated methods for using dynamic thresholding with linear robotic AR test residuals to compensate for various modeling inaccuracies [11,32]. Instead of adapting the AR tests to the nonlinear systems, this work takes the practical approach of developing a system that runs in parallel to the AR system, predicts when the modeling inaccuracies will likely be large, and increases the thresholds on the AR residual tests appropriately.

Starosweicki and Comtet-Varga have produced some interesting work describing rigorous nonlinear AR limited classes of nonlinear systems [28,29]. This work discusses several methods of rigorously developing various AR-like test residuals without actually using the nonlinear observability. This work considers the spanning issue, but is limited by its neglect of the core observability issues of AR.

Isidori and De Persis have derived a geometric residual generator using nonlinear observability [7]. These residuals are similar in concept to those generated by AR in that they use the null space of the observability to test the system behavior. However, they are limited to checking for faults where both the disturbance and fault dynamics are known well enough to model accurately, making span guarantees much more troublesome [12,25]. This limits the Isidori approach to detecting well known and well modeled faults, while AR and NLAR are geared to detect all deviations from the model.

3 NLAR Applied to a Hydraulic System

The Rosie mobile worksystem [1,5,26] discussed in this section was the initial motivation for the study of nonlinear analytical redundancy. The nonlinearities introduced by hydraulic flow through components such as servovalves are a significant issue when dealing with hydraulic robots and systems [2,6,14]. These nonlinearities led us to investigate the application of modified AR fault detection techniques to deal in an approximate manner as shown in Appendix 9 with the nonlinear hydraulic system [18–20,23]. Eventually, we left the approximation limitations of our early techniques behind with increasingly rigorous notions of nonlinear analytical redundancy.

3.1 The Rosie Mobile Worksystem

The Rosie Mobile Worksystem is a tele-robotically operated, hydraulically driven robot, which provides locomotion and a four degree-of-freedom heavy manipulator arm which can be equipped with various tools and robot manipulators. Figure 1 is a photograph of the Rosie worksystem working at the CP-5 reactor. As described in the literature [1,5,26], the robot consists of two main components or modules. The first module is a locomotor or mobile platform upon which is mounted the second module, a heavy manipulator. The locomotor module supports and transports the manipulator, and supplies it with power and control/communications. The locomotor consists of a central spine, or body core, upon which are attached front and rear drive wheel assemblies, an electronics enclosure, a hydraulic power supply system, a hydraulic enclosure for filters and valving, and a tether system. The locomotor platform is 198 cm wide, 107 cm high and 290 cm long (78 x 42 x 114 in.), supports an overall machine weight of 6,350 kg (14,000 lb.), and has a maximum speed of 0.6 m/s (2.0 ft/s).

Each wheel is powered individually by means of a geared, piston-type hydraulic motor, and is independently steered by means of a rotary actuator above that wheel. The front wheels are mounted on beams that can extend to provide additional stability when the manipulator arm is extended, as shown in Figure 2. The rear wheels are mounted on a pivoting beam for steering purposes.

The HPSS consists of a 45 kW (60 hp) supply which provides 114 l/min. (30 GPM) at 20.7 MPa (3,000 psi) for all robot operations. Electrical power and control are provided through a 61 m (200 ft) tether which is wound on a powered reel at the rear of the unit.

The heavy manipulator module supports and positions the tools that actually perform the decontamination and dismantlement (D&D) functions. Shown with a fully-extended boom in Figure 2, the heavy manipulator performs the functions of waist rotation, shoulder pitch, outer forearm extension, inner forearm extension, and wrist pitch. The hardware to execute each of these functions is very similar and consists of a flow servovalve, an actuator,

Fig. 1. The Rosie Mobile Worksystem at the CP-5 Reactor. *(Courtesy of Sharon Curd at Oak Ridge National Labs)*

and fluid components (tubing, fittings, etc.). The shoulder pitch and forearm extension functions have piston-cylinder actuators and the waist and wrist rotation are achieved through rotary actuators.

3.2 The Hydraulic Testbed and Fault Simulation

Failure modes, effects, and criticality analysis (FMECA) and fault tree based reliability analysis by researchers working with Rosie determined that the hydraulic wheel actuator subsystem was a vital component for the reliability of the mobile platform. A failure of a wheel mechanism might prevent the removal of the chassis from the reactor work site. This led to a project where the goal was to develop effective data analysis procedures (in our case AR-based fault detection) for hydraulic wheel actuators and then implement them on a testbed system located at Foster-Miller Inc., a company with considerable experience with hydraulic systems. The reliability of existing and future robots could then be enhanced by the results of this project.

Fig. 2. Rosie with manipulator arm extended. *(Courtesy of Sharon Curd at Oak Ridge National Labs)*

The testbed was a hydraulic motor controlled by a hydraulic servovalve. (This was the main nonlinear component of the system.) The specific component selected for the test program was the hydraulic wheel motor from Rosie, a Black Bruin Model 404-080-2111 from Valmet Power Transmission Inc. It has a radial piston design and has a maximum power rating of 35 kW (47 hp), a maximum output speed of 185 rpm and can deliver a torque of 2990 Nm at 250 bar (2205 ft-lb at 3600 psi). This motor was capable of driving a wheel directly and therefore accepting a substantial radial load, the exact value of load depending on the axial location of the load with respect to the motor.

The concept for the test rig itself is shown in Figures 3 and 4. A hydraulic motor powered by a HPSS was mounted on a machine bed. The output shaft was loaded radially by means of an adapter and a hydraulic jack assembly. Load was applied to the motor by means of an identical motor, used as a pump. The pump loading device was connected to the motor through a flexible coupling and differed from the motor in not having a hardened shaft and not having the "freewheeling with springs" option. The pump was fed through a separate hydraulic supply consisting of a low-pressure pump, cooler and reservoir. Load was controlled by means of a throttling valve, with a relief valve to prevent overpressure. Table 1 shows a listing of the motor-servovalve

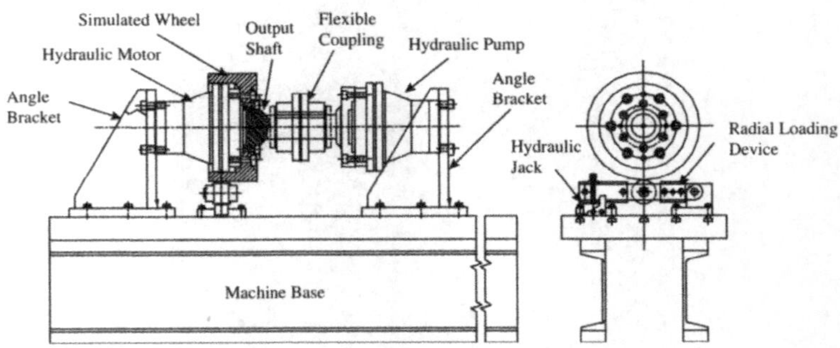

Fig. 3. Hydraulic test rig schematic. *(Courtesy of Foster-Miller)*

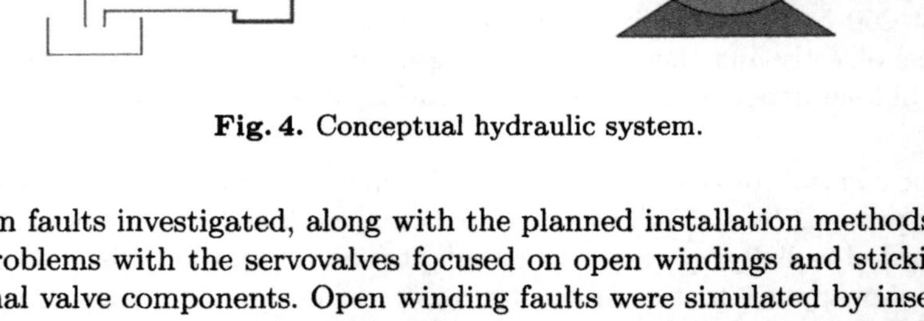

Fig. 4. Conceptual hydraulic system.

system faults investigated, along with the planned installation methods.

Problems with the servovalves focused on open windings and sticking of internal valve components. Open winding faults were simulated by inserting a relay in series with the winding. The relay was actuated by means of a bit output of the data acquisition board. This allowed a simple simulation of the fault in software.

Table 1. Faults for the Foster-Miller hydraulic testbed.

Component	Fault	Installation Method
Servovalve	Open winding	Relay
Servovalve	Sticking valve	Change control profile in software
Tachometer	Failed tachometer	Make input zero in controller
Hydraulic motor	Ruptured line	Tee flow to separate tank

A sticking valve was simulated by altering the software control profile for the valve. For example, a new open loop control profile was substituted for the standard PID closed loop control algorithm. The new algorithm incorporated stick-slip behavior as needed to simulate the sticking valve.

Hydraulic systems are clearly vulnerable to many faults that electrical systems do not experience and are much harder to model due to their inherently nonlinear nature. However, a hydraulic system has considerable advantages as an actuator in a radioactive environment, as such systems are rugged and powerful, and much less likely to produce dangerous sparks than an electrical system. Therefore, it is sensible to use one and simply put some extra time and energy into ensuring the hydraulic system is adequately monitored.

3.3 The Mathematical System Model

Begin by defining the terms used in the model.

Notation:

- A, B, and C are the canonical discrete time state-space system matrices
- B_m is the viscous damping coefficient
- $C_{tm} = c_{em} + c_{im}$ represent total, external, and internal leakage, respectively
- d_m is the volumetric displacement of the motor
- J_t is the inertia of the motor and load
- K_f,k_q and k_c are valve flow coefficients
- $M = k_c + C_{tm}$ is a generalized pressure coefficient
- p_l and $p(k)$ are the (continuous and discrete) pressure drop across the motor as measured by the sensors
- p_s is the hydraulic power supply; nominal pressure of 3000 PSI
- Q is the net fluid flow into the spool valve
- R_1 through R_5 are nonlinear AR tests
- t is the continuous time variable, k the discrete time variable, Δt is the time step
- T_g is the torque generated by the motor
- T_l is the load torque
- u_v and $u(k)$ are the (continuous and discrete) servovalve positions
- v_t is the volume of fluid within the motor
- $y(k)$ is the state vector

- β_e is the bulk modulus of the hydraulic fluid
- θ_m and $\theta(k)$ are the (continuous and discrete) position of the motor shaft - the derivative of this is measured by a velocity sensor
- ρ is the hydraulic fluid density.

The following model equations are standard for hydraulic systems [27,31]:

$$T_g = p_l d_m = J_t \ddot{\theta}_m + B_m \dot{\theta}_m + T_l, \tag{5}$$

$$Q = u_v K_f \sqrt{2(p_s - p_l)/\rho} = d_m \dot{\theta}_m + (c_{im} + c_{em}) p_l + \frac{v_t \dot{p}_l}{4\beta_e} \tag{6}$$

The state-space control model uses the following state vector:

$$\underline{x} = \begin{bmatrix} \theta \\ \dot{\theta} \\ p_l \end{bmatrix}. \tag{7}$$

The second and third state variables are instrumented, but the first (θ) is not. The nonlinear control system formed by these assumptions is as follows:

$$\dot{\underline{x}} = f(\underline{x}) + g(\underline{x})u, \tag{8}$$

$$\begin{bmatrix} \dot{\theta} \\ \ddot{\theta} \\ \dot{p}_l \end{bmatrix} = \begin{bmatrix} 0 & 1 & 0 \\ 0 & -B_m/J_t & d_m/J_t \\ 0 & -4\beta_e d_m/v_t & -4\beta_e C_{tm}/v_t \end{bmatrix} \begin{bmatrix} \theta \\ \dot{\theta} \\ p_l \end{bmatrix} + \begin{bmatrix} 0 \\ 0 \\ (4\beta_e K_f/v_t)\sqrt{2(p_s - p_l)/\rho} \end{bmatrix} u_v, \tag{9}$$

$$\underline{y} = C\underline{x}, \quad C = \begin{bmatrix} 0\,1\,0 \\ 0\,0\,1 \end{bmatrix}, \quad \underline{y} = \begin{bmatrix} \dot{\theta} \\ p_l \end{bmatrix} \tag{10}$$

The intractability of nonlinear systems such as the one described above makes fault detection techniques like AR more important. The behavior of a nonlinear system is harder to predict and control, resulting in reduced safety and reliability, which in turn makes fault detection more prominent. However, model-based fault detection techniques require good models to be effective, and traditional linear AR is generally unsatisfactory for systems with significant nonlinear components. Figure 5 shows the modeling error for a standard linearization of the Rosie wheel actuator. As this error is not accounted for in linear AR, but is modeled in the NLAR method, the figure also shows the expected improvement in going from linear to nonlinear AR.

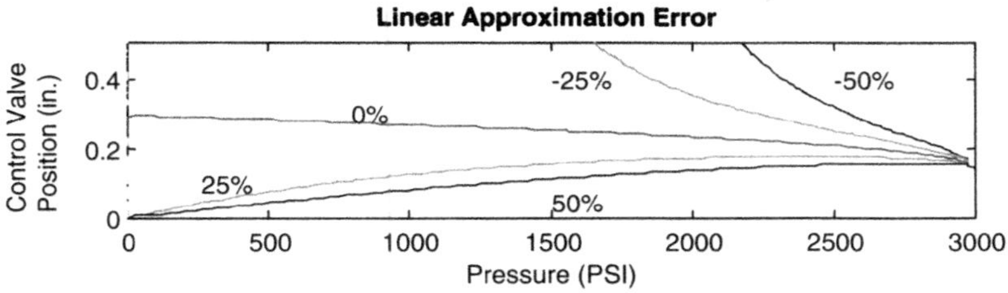

Fig. 5. Error due to linearization for hydraulic servovalve.

4 Nonlinear Analytical Redundancy for the Rosie Wheel Actuator

Now we apply our nonlinear analytical redundancy (NLAR) technique to the Rosie subsystem described above. This will allow us to avoid the mathematical errors caused by linearization, as shown in Figure 5. To begin, we calculate the grouped observability vector $\mathcal{O}_\triangle$:

$$\mathcal{O}_\triangle = \begin{bmatrix} c_1\underline{x} \\ L_f c_1 \underline{x} + L_g c_1 \underline{x} u \\ L_{ff} c_1 \underline{x} + L_{fg} c_1 \underline{x} u + L_{gf} c_1 \underline{x} u + L_{gg} c_1 \underline{x} u^2 \\ c_2\underline{x} \\ L_f c_2 \underline{x} + L_g c_2 \underline{x} u \\ L_{ff} c_2 \underline{x} + L_{fg} c_2 \underline{x} u + L_{gf} c_2 \underline{x} u + L_{gg} c_2 \underline{x} u^2 \end{bmatrix},$$

$$\mathcal{O}_\triangle = \begin{bmatrix} \dot{\theta} \\ -B_m\dot{\theta}/J_t + d_m p_l/J_t \\ \left\{ \begin{array}{l} \left(B_m^2/J_t^2 + -4\beta_e d_m^2/J_t v_t\right)\dot{\theta} \\ + \left(-B_m d_m/J_t^2 + -4\beta_e d_m C_{tm}/J_t v_t\right) p_l \\ + \left(4\beta_e d_m K_f/J_t v_t\right)\sqrt{2(p_s - p_l)/\rho}\, u \end{array} \right\} \\ p_l \\ \left\{ \begin{array}{l} (-4\beta_e d_m/v_t)\dot{\theta} + (-4\beta_e C_{tm}/v_t) p_l \\ +(4\beta_e K_f/v_t)\left(\sqrt{2(p_s - p_l)/\rho}\right) u \end{array} \right\} \\ \left\{ \begin{array}{l} \left(\begin{array}{l} 4B_m\beta_e d_m/J_t v_t + 16\beta_e^2 d_m C_{tm}/v_t^2 \\ + \left(8\beta_e^2 d_m K_f/v_t^2\right)\sqrt{2/\rho(p_s - p_l)}\, u \end{array} \right)\dot{\theta} \\ + \left(\begin{array}{l} -4\beta_e d_m^2/J_t v_t + 16\beta_e^2 C_{tm}^2/v_t^2 \\ + \left(8\beta_e^2 K_f C_{tm}/v_t^2\right)\sqrt{2/\rho(p_s - p_l)\, u} \end{array} \right) p_l \\ (-16\beta_e^2 K_f C_{tm}/v_t^2)\left(\sqrt{2(p_s - p_l)/\rho}\right) u \\ + \left(-16\beta_e^2 K_f^2/\rho v_t^2\right) u^2 \end{array} \right\} \end{bmatrix}. \quad (11)$$

Note the nonlinear sub-observability matrices $c_i \nabla \mathcal{O}_\triangle$ have no terms containing θ, so the system is rank two and only three terms are needed in each C_j. This also means four independent NLAR tests are expected.

Now calculate the null-matrix Ω:

$$\Omega \mathcal{O}_\triangle = \begin{bmatrix} \Omega_{11} \; \dot{\theta} \; 0\,0\,0\,0 \\ \Omega_{21} \; 0 \; \dot{\theta} \; 0\,0\,0 \\ \Omega_{31} \; 0\,0 \; \dot{\theta} \; 0\,0 \\ \Omega_{41} \; 0\,0\,0 \; \dot{\theta} \; 0 \\ \Omega_{51} \; 0\,0\,0\,0 \; \dot{\theta} \end{bmatrix} \mathcal{O}_\triangle = 0\,, \tag{12}$$

where:

$$\Omega_{11} = \frac{B_m}{J_t}\dot{\theta} + \frac{-d_m}{J_t} p_l,$$

and

$$\Omega_{12} = \begin{matrix} \left(\frac{-B_m^2}{J_t^2} + \frac{4\beta_e d_m^2}{J_t v_t}\right)\dot{\theta} + \left(\frac{B_m d_m}{J_t^2} + \frac{4\beta_e d_m C_{tm}}{J_t v_t}\right) p_l \\ + \left(\frac{-4\beta_e d_m K_f}{J_t v_t}\sqrt{2(p_s - p_l)/\rho}\right) u \end{matrix},$$

$$\Omega_{13} = -p_l,$$

$$\Omega_{14} = \begin{matrix} 4\beta_e d_m / v_t \dot{\theta} + 4\beta_e C_{tm}/v_t p_l \\ + (-4\beta_e K_f / v_t)\sqrt{2(p_s - p_l)/\rho}\; u \end{matrix},$$

$$\Omega_{51} = \begin{matrix} \begin{pmatrix} -4B_m\beta_e d_m / J_t v_t - 16\beta_e^2 d_m C_{tm}/v_t^2 \\ + \left(-8\beta_e^2 d_m K_f / v_t^2\right)\sqrt{2/\rho(p_s - p_l)\,u} \end{pmatrix}\dot{\theta} \\ + \begin{pmatrix} 4\beta_e d_m^2 / J_t v_t - 16\beta_e^2 C_{tm}^2 / v_t^2 + \\ \left(-8\beta_e^2 K_f C_{tm}/v_t^2\right)\sqrt{2/\rho(p_s - p_l)\,u} \end{pmatrix} p_l \\ + \left(16\beta_e^2 K_f C_{tm}/v_t^2\right)\sqrt{2(p_s - p_l)/\rho}\, u + 16\beta_e^2 K_f^2/\rho v_t^2 u^2 \end{matrix}.$$

Then calculate $\mathcal{O}_{\triangle DD}$:

$$\mathcal{O}_{\triangle DD} = \begin{bmatrix} y_1(t) \\ \dot{y}_1(t) \\ \ddot{y}_1(t) \\ y_2(t) \\ \dot{y}_2(t) \\ \ddot{y}_2(t) - (4\beta_e K_f / v_t)\sqrt{2(p_s - y_2(t))/\rho}\; \dot{u}(t) \end{bmatrix}. \tag{13}$$

Then simply apply the NLAR equation:

$$\begin{bmatrix} \Omega_{11} \; \dot{\theta} \; 0\,0\,0\,0 \\ \Omega_{21} \; 0 \; \dot{\theta} \; 0\,0\,0 \\ \Omega_{31} \; 0\,0 \; \dot{\theta} \; 0\,0 \\ \Omega_{41} \; 0\,0\,0 \; \dot{\theta} \; 0 \\ \Omega_{51} \; 0\,0\,0\,0 \; \dot{\theta} \end{bmatrix} \begin{bmatrix} y_1(t) \\ \dot{y}_1(t) \\ \ddot{y}_1(t) \\ y_2(t) \\ \dot{y}_2(t) \\ \ddot{y}_2(t) - (4\beta_e K_f / v_t)\sqrt{2(p_s - y_2(t))/\rho}\; \dot{u}(t) \end{bmatrix} = 0. \tag{14}$$

to get the NLAR tests.

$$\boxed{R_1 = -\dot{y}_1(t) - \frac{B_m}{J_t} y_1(t) + \frac{d_m}{J_t} y_2(t)} \tag{15}$$

$$\boxed{R_2 = \begin{array}{l} -\ddot{y}_1(t) + \left(B_m^2/J_t^2 - 4\beta_e d_m^2/J_t v_t\right) y_1(t) \\ + \left(-B_m d_m/J_t^2 - 4\beta_e d_m C_{tm}/J_t v_t\right) y_2(t) \end{array}} \tag{16}$$

$$\boxed{R_3 = -y_2(t) y_1(t) + y_1(t) y_2(t)} \tag{17}$$

$$\boxed{R_4 = \begin{array}{l} -\dot{y}_2(t) - 4\beta_e d_m y_1(t)/v_t - 4\beta_e C_{tm} y_2(t)/v_t \\ + (4\beta_e K_f/v_t) \sqrt{2(p_s - y_2(t))/\rho}\, u \end{array}} \tag{18}$$

$$\boxed{R_5 = \begin{array}{l} -\ddot{y}_2(t) \\ + \left(\begin{array}{l} 4B_m\beta_e d_m/J_t v_t + 16\beta_e^2 d_m C_{tm}/v_t^2 \\ + \left(8\beta_e^2 d_m K_f/v_t^2\right) \sqrt{2(p_s - y_2(t))/\rho}\, u \end{array} \right) y_1(t) \\ + \left(\begin{array}{l} -4\beta_e d_m^2/J_t v_t + 16\beta_e^2 C_{tm}^2/v_t^2 \\ + \left(8\beta_e^2 K_f C_{tm}/v_t^2\right) \sqrt{2(p_s - y_2(t))/\rho}\, u \end{array} \right) y_2(t) \\ - \left(16\beta_e^2 K_f C_{tm}/v_t^2\right) \sqrt{2(p_s - y_2(t))/\rho}\, u - 16\beta_e^2 K_f^2 u^2/\rho v_t^2 \\ + (4\beta_e K_f/v_t) \sqrt{2(p_s - y_2(t))/\rho}\, \dot{u} \end{array}} \tag{19}$$

Note that R_3 is trivial, so the expected number of independent NLAR tests is generated. R_1 and R_4 correspond to the model equations. R_2 and R_5 correspond to the convolved first derivatives of the model equations. This is a common result of AR and NLAR analyses, and a reassuring one. The tests generated relate directly and intuitively to the system being analyzed.

5 Results

The following results were generated by applying nonlinear analytical redundancy techniques to recorded experimental data from the hydraulic testbed described previously. In the case of the Rosie wheel actuator, the NNAR[1] tests discussed in Appendix 9 and in previous papers [18–20,23] are the same

[1] It turns out that the spans of linear and nonlinear observability spaces for this sytem are similar enough that previously derived approximate techniques [19,23], created by us to allow application of linear AR to nonlinear systems, were correct for this particular system. (This is something of a co-incidence, and not true for a generaleralized nonlinear system.) For this specific case the approximation used in the NNAR technique is exact. The nearly nonlinear AR (NNAR) technique

as the NLAR tests. While this is a useful coincidence validating previously derived approximate test results, the most important aspect of the following results remains that they validate the NLAR technique using physical testbed data.

Each of the following result sections shows all four of the NLAR test residuals for both a baseline data run (light grey) and a run with an injected fault (black). As NLAR test residual magnitudes are a somewhat arbitrary function of basis choice during derivation, it is important to note that the magitude of the plot along the y-axis is important only in comparison with other runs. It is thus the shape and magnitude of the residual *relative to the baseline test run* that is indicative of faulty or fault-free conditions.

The hydraulic engine speed and the load on the system (in terms of PSI across the load pump) are shown with each set of residuals.

5.1 Servovalve, Open Winding Fault

This fault was modeled by physically opening the winding on the hydraulic servovalve and inserting a relay to short circuit the control solenoid. The fault is detected easily by NLAR -the onset and duration of this fault is clear in all AR tests - large spikes in two of the NLAR test residuals (R_1 and R_5) show the fault the instant after it appears. Large steps in the other two tests are almost as fast. This is because the open winding acts like a large step input that is not accounted for in the model, degrading the accuracy of the model greatly in a short time period. This naturally provokes a strong AR response, which allows us to determine the system is operating under faulty conditions.

5.2 Sticking Wheel Motor Servovalve Fault

In the simulation of this fault, the system was ramped from zero speed to a speed of 5 RPM (equal to the fault free run) while the control input was intermittantly set to zero to simulate sticking. This fault is evident on all of the AR tests, although R_1 and R_5 show the clearest results. The system starts out close to the model but rapidly departs from model-following behavior if the control system tries to apply a nonzero control input while the servovalve is 'stuck.' This causes the spikes in the NLAR tests shown in Figure 5.2 that make the fault easily detectable.

---

is an approximate but useful method developed early in the course of this work to deal with the nonlinearities of the Rosie hydraulic actuator [18–20,23]. This method is not truly nonlinear in nature, however, as it uses the linearized system model to derive the AR tests. (The approximate NNAR and PLAR methods are briefly described in Appendix 9.)

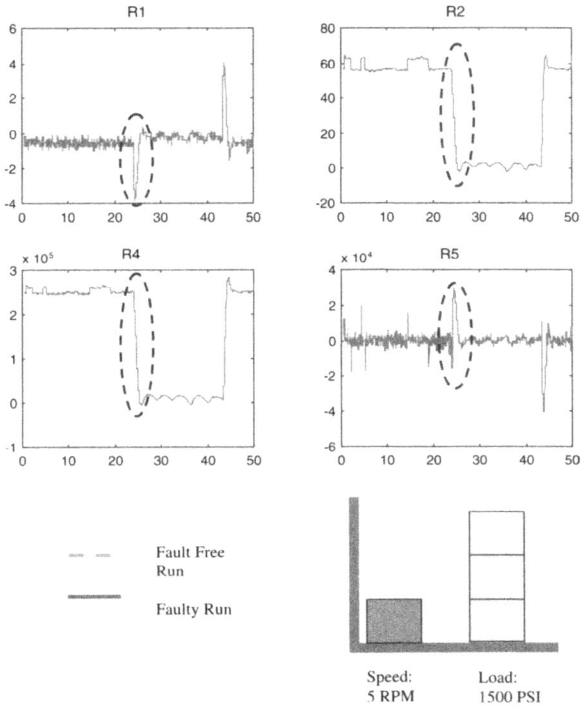

Fig. 6. Servovalve, open winding fault run.

5.3 Loss of Speed Feedback (Tachometer) Fault

The failed sensor invalidates the control loop if the commanded velocity is nonzero. The sensor reports zero angular velocity, so the controller tries to increase the speed by providing more power, quickly leading to a runaway system. (The testbed had a limiting device to prevent damage when this happened.) NLAR detects this as a deviation from the model-expected behavior almost instantly - the control inputs and sensor values disagree. This leads to the clear error signals seen in the figure.

6 The Importance of Model Accuracy

It bears repeating that the success of the NLAR fault detection technique is limited by the accuracy of the system model. The successful tests described

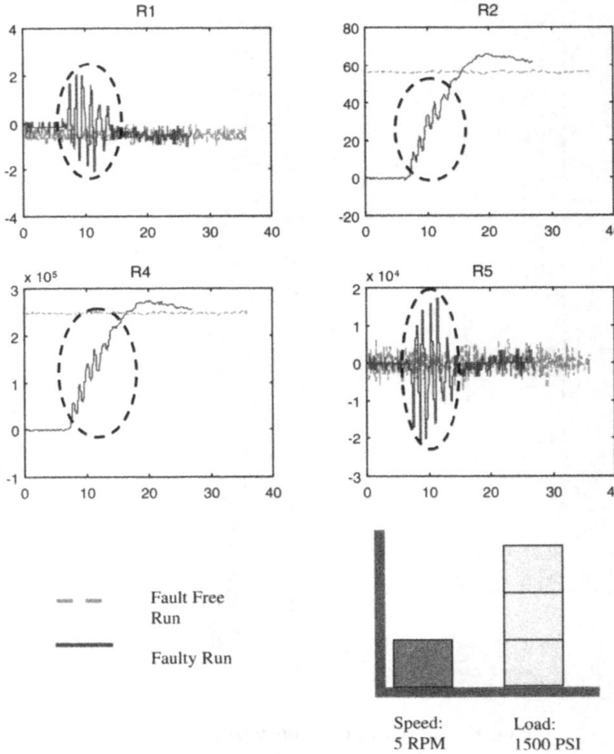

Fig. 7. Sticking wheel motor servovalve fault run.

in Section 5 are notable in that they are detecting faults that cause large and immediate deviations from the system model described in Section 3.3. All faults are not so tractable. Consider the following fault, which represents a leak in the motor-valve system, shown in figure 9.

Why does NLAR not generate a clear and unambiguous signal as it did for the other three faults? Investigation shows that as NLAR is a model based fault detection algorithm, it is limited by the accuracy of the model. The NLAR test residuals shown were applied as a post-analysis to an incompletely documented system. While the model presented in Section 3.3 is a good mathematical representation of this hydraulic system, several of the constants in the system are only approximately determined or not as constant as one might desire. For example, the supply pressure p_s is approximated as

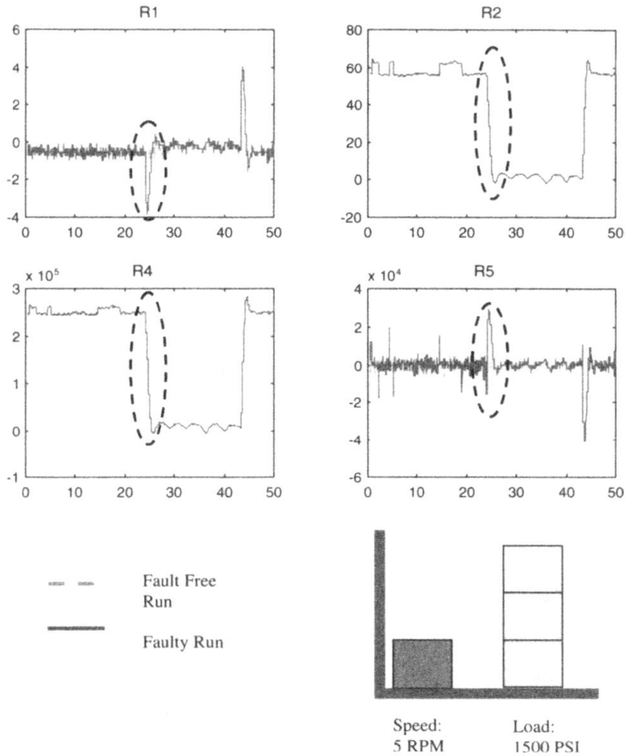

Fig. 8. Loss of speed feedback (tachometer) fault run.

a constant 3000 PSI. This is a rough approximation made necessary by the lack of a pressure sensor to give this value directly. Similarly, many of the parameters describing the liquid behavior of the hydraulic fluid are actually functions of the oil temperature. Several other constants are catalog values of uncertain accuracy. For most error based control applications, these approximations are acceptable. However, for model based fault detection, every inaccuracy limits the resolution of the tests. In the case of the leak fault shown, the resolution of the NLAR tests is too coarse to detect the tiny error caused by the steady leakage of a small percentage of the hydraulic fluid. We are confident that better modeling and proper instrumentation of the system could overcome this and make NLAR capable of detecting quite such leak faults. For many sensitive systems, better modeling is a small price to pay for improved reliability, and NLAR is a good choice for such systems.

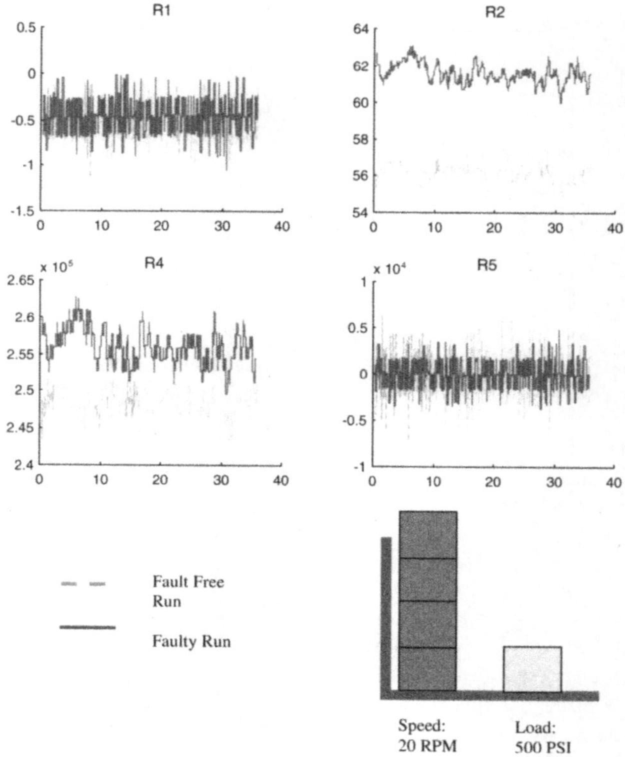

Fig. 9. Leak in motor-valve system fault.

7 Conclusions

Analytical redundancy is a model based fault detection technique, and thus requires an accurate system model to be effective. Previous AR theory has been limited to linear system models; this is a serious problem in a nonlinear world. Advances in nonlinear control theory have opened the door to the nonlinear realm. This allows us to apply the AR method rigorously to most nonlinear systems without losing any of the valuable theoretical assurances of complete coverage inherent to the AR method. The Rice and Clemson group has developed such theoretically robust nonlinear AR techniques.

Nonlinear AR fault detection is a useful monitoring method for hydraulic systems such as Rosie that must operate in hazardous environments. Better fault detection for hydraulics will reduce the costs associated with failures of such systems in the workplace by minimizing damage done by and to faulty systems as well as the amount of time wasted by false alarms. Safety and reliability are critical for success of many operations in hazardous environments, and this work represents the first detailed examination of nonlinear AR fault detection for these types of hydraulic robot systems.

8 Acknowledgements:

This work was supported in part by the National Science Foundation under grants IRI-9526363 and CMS 9796328, by DOE Sandia National Laboratory Contract #AL3017, NASA grant NAG5-9785, NSF/EPSCoR grant EPS-9630167, and DOE contract DE-FG07-97ER 14830. The authors would also like to thank V. Jammu, R. W. Gamache, M. Martin, T. Walter, and D. McCauley at Foster-Miller Technologies, Albany NY

9 Appendix: Approximate Approaches to NLAR

One way to deal with a system model that is nonlinear enough to change considerably as it moves through the workspace is to create several sets of AR tests for the system linearized about state vectors located in each region of interest. In the local region each set would be more accurate than a general linearization of the control equations for the entire workspace. An example of this technique, and the improvement it brings, is illustrated below in Figure 10. In this case, generated for the hydraulic servovalve system, the pressure-valve position workspace in which the flow equation is nonlinear is divided up into nine equal regions. The model equation is linearized about a point at the center of each and normal AR tests are derived. (Due to the symmetry of the system, only four linearizations are needed in practice.) During operation, the AR test used is the one that was linearized about a point closest to the current position, with interpolated transitions near the borders of each region.

Figure 11 shows the results from a fault free run and a faulty run (the fault was a large leak) of PLAR tests on a simulated hydraulic servovalve [18,19]. The fault free run shows how PLAR minimizes the drifting away from the model errors of pure linear AR. Before the test residual can drift far from the point about which it was linearized, the system transitions into another, more appropriate AR test linearized about a point closer to the actual state of the system, leading to a saw-toothed residual about the nominally correct zero value. The results in a test run with a large leak fault added to the hydraulic system show that this saw-tooth is about an order of magnitude smaller than the fault signature.

Nearly nonlinear analytical redundancy is a natural outgrowth of PLAR. Although each PLAR partition uses a different linearization of the control equations, the different linearizations all share the same basic form - they

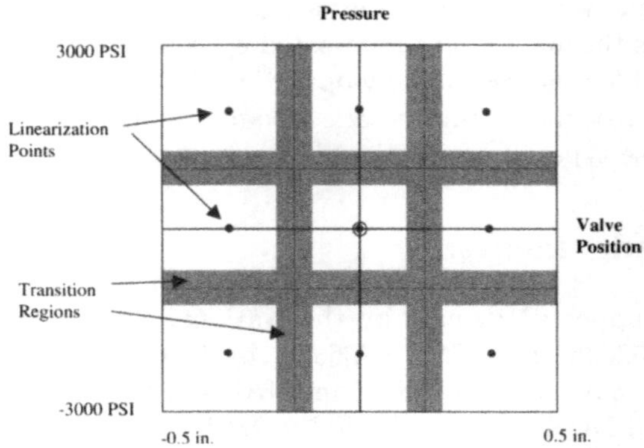

Fig. 10. PLAR division of hydraulic servovalve workspace.

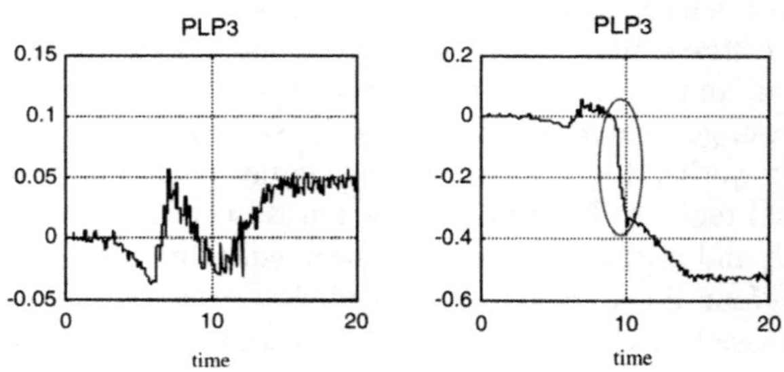

Fig. 11. PLAR fault free (left) and faulty (right) test results for simulated hydraulic servovalve.

are tangents to the control system at the point of linearization. This means that the AR tests generated by the different linearizations will be, in essence, tangents to some more accurate nonlinear AR test. It is reasonable to approach this test by dividing the workspace into many closely packed regions and taking appropriate linearizations. In the limit of infinitely small regions, this is a nonlinear AR test! However, there is a much simpler method of finding these tests. Recall that AR tests tend to be combinations of the model equations, sensor comparisons, and their derivatives. By performing linear AR on a linearized system and identifying the relationship between the AR residual tests and the control model and sensors, it is possible to find nonlinear AR tests by simply duplicating this relationship to the nonlinear control system and sensors! For example, if the AR tests for the linear system are the linearized model equations, their first derivatives, and direct sensor comparisons between the linearized sensors, the nonlinear model equations, their first derivatives, and comparisons of the nonlinear sensors can be used as NNAR tests.

Why, then, are these tests referred to as merely "nearly nonlinear" rather than "fully nonlinear?" Unfortunately, as they use the linear observability matrix they are still using linear approximations of the observability null-space, Ω. It can be shown that this is different than the nonlinear null-space; in fact the two can have a different rank. The AR guarantees of testing for all of the possible model deviations efficiently are thus not valid in PLAR and NNAR. *Any* AR method that doesn't use the full nonlinear observability space will suffer from this drawback. This is what makes fully nonlinear AR using the nonlinear observability space so desirable, and relegates PLAR and NNAR to secondary roles. These methods do not require extensive nonlinear calculation to use, but they are essentially approximations of the complete NLAR technique.

References

1. Bares L. C., Conley L.S., Thompson B.R. (1997) Rosie: A Mobile Worksystem for D & D: Overview of System Capabilities and CP-5 Reactor Application. Proc. of ANS 7th Topical Meeting on Robotics and Remote Systems, Augusta, GA, 471–477
2. Bu F., Yao B. (2001) Nonlinear Model Based Coordinated Adaptive Robust Control of Electro-hydraulic Robotic Arms via Overparametrizing Method. IEEE International Conference on Robotics and Automation, Seoul, Korea
3. Caccavale F., Walker I.D. (1997) Observer-Based Fault Detection for Robot Manipulators. IEEE Int. Conf. on Robotics and Automation. Albuquerque, NM, 2881–2887
4. Chow E.Y., Willsky A.S. (1984) Analytical Redundancy and the Design of Robust Failure Detection Systems. IEEE Trans. on Automatic Control 29:603–614
5. Conley L., Hamel W.R., Thompson B.R. (1995) Rosie: A Mobile Worksystem for Decontamination and Dismantlement Operations. Proc. of the ANS 6th Topical Meeting on Robotics and Remote Systems, Monterey, CA, 231–238
6. Daachi B., Benallegue A., M'Sirdi N. K. (2001) A Stable Neural Adaptive Force Controller for a Hydraulic Actuator. IEEE Int. Conf. on Robotics and Automation, Seoul, Korea
7. De Parsis C., Isidori A. (2001) A Geometric Approach to Nonlinear Fault Detection and Isolation. IEEE Trans. on Automatic Control 46:853–865
8. Department of Energy, Washington, D.C. (1993) Environmental Restoration and Waste Management 5-Year Plan, Fiscal Years 1994-1998. DOE/S-00097P, Vol. 1-2

9. Department of Energy, Federal Energy Technology Center, Morgantown, WV (1997) Environmental Waste Management Project Fact Sheet. http://www.fetc.doe.gov/publications/factsheets/ewm/index.html
10. Dhillon B.S. (1991) Robot Reliability and Safety, Springer-Verlag, New York, NY
11. Ding X., Frank P.M. (1991) Frequency Domain Approach and Threshold Selector for Robust Model-Based Fault Detection and Isolation. Proc. of IFAC Fault Detection, Supervision and Safety for Technical Processes, Baden-Baden, Germany, 271–276
12. Ding X., Guo L., Jeinsch T. (1999) A Characterization of Parity Space and Its Application to Robust Fault Detection. IEEE Transactions on Automatic Control 44:337–343
13. Hammouri H., Kinnaert M., El Yaagoubi E.H. (1999) Observer-Based Approach to Fault Detection and Isolation for Nonlinear Systems. IEEE Transactions on Automatic Control 44:1879–1884
14. Honegger M., Corke P. (2001) Model-Based Control of Hydraulically Actuated Manipulators. IEEE International Conference on Robotics and Automation, Seoul, Korea
15. Isidori A. (1995) Nonlinear Control Systems. Springer-Verlag, London,UK
16. Leuschen M.L. (2001) Derivation and Application of Nonlinear Analytical Redundancy Techniques with Applications to Robotics. PhD thesis, Department of Electrical and Computer Engineering, Rice University, Houston, TX
17. Leuschen M.L., Walker I.D., Cavallaro J.R. (1998) Robot Reliability Through Fuzzy Markov Models. Proc. IEEE Annual Reliability and Maintainability Symposium, Anaheim, CA, 209–214
18. Leuschen M.L., Walker I.D., Cavallaro J.R. (1999) Investigation of Reliability for Hydraulic Robots Using Analytical Redundancy. Proc. IEEE Annual Reliability and Maintainability Symposium, Washington, DC, 122–128
19. Leuschen M.L., Walker I.D., Cavallaro J.R. (1999) Monitoring and Diagnostics for a Hydraulic Robot in Hazardous Environments. Proc. of ANS 8th Topical Meeting on Robotics & Remote Systems, Pittsburgh, PA
20. Leuschen M.L., Walker I.D., Cavallaro J.R. (2001) Experimental AR Fault Detection Methods for a Hydraulic Robot. Proc. of ANS 9th Topical Meeting on Robotics and Remote Systems, Seattle, WA, F-131
21. Leuschen M.L., Walker I.D., Cavallaro J.R. (2002) Robotic Fault Detection Using Nonlinear Analytical Redundancy. IEEE International Conference on Robotics and Automation, Washington, DC
22. Leuschen M.L., Walker I.D., Cavallaro J.R. (2002) Nonlinear Analytical Redundancy for Fault Detection. Submitted to: IEEE Trans. on Automatic Control
23. Leuschen M.L., Walker I.D., Cavallaro J.D., Gamache R.W., Martin M. (2000) Experimental AR Fault Detection Methods for a Hydraulic Robot. Proc. of 18th International Systems Safety Conference, Fort Worth, TX, 402–409
24. Magni J.F., Mouyon P. (1994) On Residual Generation by Observer and Parity Space Approaches. IEEE Trans. on Automatic Control 39:441–447
25. Nyberg M., Nielsen L. (2000) A Universal Chow-Willsky Scheme and Detectability Criteria. IEEE Trans. on Automatic Control 45:152–156
26. RedZone Robotics Inc. The Rosie Mobile Worksystem. http://www.redzone.com/downloads/Rosie.pdf.
27. Stadler W. (1995) Analytical Robotics and Mechatronics. Mc-Graw-Hill, Inc., New York, NY

28. Staroswiecki M., Cassar J.P, Comtet-Varga G. (1997) Analytic Redundancy Relations for State Affine Systems. Proc. of 1997 Fourth European Control Conference, Brussels, Belgium
29. Staroswiecki M., Comtet-Varga G. (2001) Analytical Redundancy Relations for Fault Detection and Isolation in Algebraic Dynamic Systems. Automatica 37:687–699
30. Visinsky M.L., Cavallaro J.R., Walker I.D. (1995) A Dynamic Fault Tolerance Framework for Remote Robots. IEEE Transactions on Robotics and Automation 11:477–490
31. White F.M. (1994) Fluid Mechanics. McGraw-Hill, Inc., USA
32. Wünnenberg J., Frank P.M. (1990) Dynamic Model Based Incipient Fault Detection Concept for Robots. Proc. of 11th IFAC World Congress, Tallinn, Estonia, 61–66
33. Zhirabok A.N., Preobragenskaya O.V. (1993) Instrument Fault Detection in Nonlinear Dynamic Systems. Proc. of 1993 IEEE International Conference on Systems, Man, and Cybernetics, Le Touquet, France, 114–119

GPSR Compliance

The European Union's (EU) General Product Safety Regulation (GPSR) is a set of rules that requires consumer products to be safe and our obligations to ensure this.

If you have any concerns about our products, you can contact us on ProductSafety@springernature.com

In case Publisher is established outside the EU, the EU authorized representative is:

Springer Nature Customer Service Center GmbH
Europaplatz 3
69115 Heidelberg, Germany

Batch number: 10584448

Printed by Printforce, the Netherlands